聚 · 世界行业精英
Gather- Global Real Estate Elites

鉴 · 全球设计之美
Select- Best Global Design

錾 · 年度荣誉之巅
Elaborate – Top Annual Award

* 本书中所有项目图片、文字信息均由项目业主单位、设计单位提供。

地建师、CREDAWARD 地产设计大奖·中国组委会保留所有权。未经版权所有人书面同意，不得以任何形式（包括但不限于电子或实体）对本出版物的任何内容进行复制及转载。

声明：本书中项目名称或公司名称可能是商标或注册商标，仅作标识解释之用，无意侵权。

*All the project pictures and the text information in this book are originated from the project owners and the design units who have authorized the publisher to use and publish them.

Original edition published by DJSER and CREDAWARD Committee. All rights reserve. No part of this publication may be reproduced or distributed by any means or stored in a database or retrieval system, without the prior written permission of the publisher.

Disclaimer: The project names or the company names in this book may be trademarks or registered trademarks, which are only used for logo interpretation and no infringement is intended.

编 委
EDITORIAL COMMITTEE

地建师
DJSER
地产设计大奖·中国 组委会
*CRED*AWARD Committee

编委会主任
DIRECTOR OF THE EDITORIAL BOARD

邢同和
Tonghe Xing

编委会成员
EDITORIAL BOARD

陈建邦　范逸汀　方　芳　冯腾飞
贾朝晖　戎武杰　阴　杰　张兆强
朱晓涓　钟　琴

K.B. Albert Chan · Yiting Fan ·
Fang Fang · Tengfei Feng ·
Zhaohui Jia · Wujie Rong ·
Jerry Yin · Zhaoqiang Zhang ·
Xiaojuan Zhu · Qin Zhong

前言

2014年全球经济缓慢复苏,作为国民经济支柱产业之一的房地产行业宣告进入白银时代,这一年是房地产史上重要的一年,更是产品力时代的伊始。遵循基本商业逻辑的设计创新和设计品质提升,预示着地产行业进入了一个新的发展阶段。在这一时代大背景下,「CREDAWARD 地产设计大奖·中国」应运而生。

「CREDAWARD 地产设计大奖·中国」,由TOP50开发商总建筑师和具有影响力的学术代表联袂担任评委,坚持"建筑审美、环境优先、用户体验、经济效益"16字评价标准,表彰全球范围内的优秀项目,传播前沿设计理念,引领未来城市发展之路,促进社会可持续发展!

从第一届到第五届,组委会共邀请到14个国家和地区的全球知名设计机构及中国地区品牌开发商积极参与,累计收到4 276份优秀作品,产生85个金奖项目。项目类型涵盖居住项目、商业项目、办公项目、酒店项目、文化项目、旅游项目、医养项目、产业项目、城市再生项目、新业态等。设计类型包含规划设计、建筑设计、景观设计、室内设计、照明设计。

源于中国、面向世界的「CREDAWARD 地产设计大奖·中国」,具有其特殊的时代性、全球性和使命感,因其严谨的评选流程、公开的评选模式、严苛的获奖机制,被行业誉为"最难获得的大奖"之一。

地建师(DJSER)作为「CREDAWARD 地产设计大奖·中国」的组织方,代表全体评委感谢众多海内外优秀项目倾情参与,毫无保留,同台竞技,共同促进社会进步发展。感谢每一家参赛机构用自己的实际行动来证明,作为社会发展的重要参与者,我们恪守对原则的尊重,捍卫设计的灵魂。

2020年是人类历史上不平凡的一年。图书作为一个时代或者一个过程的记录载体,具有传承文化和连接沟通纽带的重要意义。这一年,地建师携手「CREDAWARD 地产设计大奖·中国」评委共同策划出版《中国地产设计集萃 I(2014—2019)》,邀请历年参赛业主及设计方对124个来自全球的第一届至第五届金、银奖作品进行整合梳理。点滴图文精酌,以书籍形式留存设计精华;坚守初心,记录下城市发展的足迹。

CREDAWARD 地产设计大奖·中国 组委会
2020年5月

FOREWORD

With the global economy slowly recovering in 2014, the real estate industry, one of the pillar industries of the national economy, announced that it has entered the silver era, meanwhile the historical year of 2014 marked the advent of product power. On the other hand, the innovative and outstanding design following the basic business logic indicates that the real estate industry has entered a new stage of development. Based on the background mentioned above, CREDAWARD, China Real Estate Design Award, came into being.

Based on the evaluation principle: Architectural Aesthetics, Environment Priority, User Experience, Economic Benefits, CREDAWARD attracted the chief architects from the Top50 developers and the influential academic representatives to serve as its jurys to reward the world-wide outstanding projects, spread the cutting-edge design concepts, lead the future urban development and facilitate the sustainable social development.

From the first session to the fifth, the organizing committee invited the world-renowned design institutions and Chinese brand developers from China and the other 14 countries and regions, received a total of 4,276 outstanding works, and produced 85 golden awards covering varieties of projects and designs, such as, residential project, business project, office project, hotel project, cultural project, tourism project, medical care project, industry project, urban regeneration project, new business form project and the planning design, architectural design, landscape design, interior design, lighting design.

With the characteristics of the times, globalization and mission, the CREDAWARD originated in China to embrace the world. It is known as one of the most difficult awards to earn in the real estate industry due to its rigorous selection process, open selection mode, and strict award mechanism.

On behalf of all the jurys, the architects, as the organizer of CREDAWARD, DJSER would like to thank all the domestic and international contestants as their actively participating, unreservedly competing on the same stage, and jointly promoting social development. Meanwhile the architects would also like to thank all the participating organizations for using their own actions to prove that they, as an important part of social development, will stick to respecting the principles and defending the soul of design.

As a recording carrier of an era or a certain process, books can inherit culture and offer the communicating platform for human beings. In the unusual year of 2020, DJSER together with the committee of CREDAWARD, planned to publish the book, Best Works of CREDAWARD I (2014-2019), to keep the essence of design in the form of books and adhere to the original intention as well as record the footprint of urban development, thus inviting the owners and designers from the participating institutions over the years to comment on the 124 gold and silver awards and select the pictures and texts from the first to the fifth sessions to make this book.

CREDAWARD Committee
2020.05

| 序 言 |

《中国地产设计集萃 I（2014—2019）》在 2021 年正式出版发行，这本书是「CREDAWARD 地产设计大奖·中国」走过五年的路，也是地产行业发展中评选出的精品汇聚，极具行业价值和社会意义。

本书收集了第一至第五届金、银奖作品。项目体现了时代进程发展的特殊性、类型的多元多样性、参与评奖项目的广泛性。

「CREDAWARD 地产设计大奖·中国」第一届执行委员会作为本书编委，对 124 个金、银奖作品反复精酌，选出 34 个可代表中国地产这些年发展的重要标杆项目，进行深度解读。编委会在梳理历年奖项申报资料的过程中，深感每一份参赛作品提交材料极其认真详尽。每届金、银奖的作品都是国内外建筑、规划、环境、室内设计、城市更新等行业（包括房产领域的著名公司）所精心打造的原创品牌，独具创新技术亮点，体现了获奖作品的含金量和影响力。我们不愿意看到如此珍贵的材料在历史潮流中逐渐丢失或被人遗忘。

在此，特别要感谢一起走过这五年的评审委员会各位评委，大家坚持正确的评审方向，正能量地推进地产发展，秉持公平、公正、理性的原则，谨对这种认真负责的精神和不辞辛劳的工作态度表示深深的感谢！致敬！

《中国地产设计集萃 I（2014—2019）》记录的中国地产这五年的设计进步和贡献，都值得受到鼓励和赞誉！也将成为"人民城市人民建，人民城市为人民"的一段辉煌成就。让我们共同见证在中国共产党领导下，中国城市化进程前所未有的高速发展。

值此建党 100 周年大庆之际，有幸将此书作为一份献礼！

邢同和

PREFACE I

Professor Tonghe Xing

With high social and industrial value, the book, Best Works of CREDAWARD I (2014-2019), which will be officially published and released in 2021, not only marks the five-year journey of CREDAWARD, but also displays the best selected works in real estate development industry.

This book collects Gold and Silver Awards of CREDAWARD from the first session to the fifth, presenting its works' uniqueness, diversity and extensiveness.

On one hand, the first executive committee of CREDAWARD, as the book's editorial board, were impressed by every submitted work in the process of sorting out the application materials for CREDAWARD over the years, on the other hand, after considering 124 Gold and Silver Awards, they selected 34 benchmark works to be elaborated, which represent the development of China real estate industry in recent years. In addition, whether Gold Awards or Silver Awards throughout the five sessions, each work, created by world-wide well-known companies in the field of architecture, planning, environment, interior design, urban renewal and real estate, is featured with high technology and innovation and original brand to reflect its value and influence. It is those precious works that write history and carry it on from generation to generation.

I wish to take this opportunity to convey my sincere thanks to all the jurys who worked with me in the past five years for their adhering to the right evaluation direction and the principle of being fair, just and rational to positively facilitate the development of real estate industry, for their serious and responsible spirit, for their tireless work attitude!

The book, Best Works of CREDAWARD I (2014-2019), records the five-year design progress and contributions in China real estate industry, which deserves encouragement and praise! This book also displays a remarkable journey of "People's cities are built by people, People's cities serve for people". Let us see through the unprecedentedly rapid development of urbanization in China under the leadership of the Communist Party of China.

On the occasion of the 100th anniversary of the founding of the Communist Party of China, I would like to take this honor to present this book as a birthday gift to the CPC!

| 序 言 ||

作为「CREDAWARD 地产设计大奖·中国」的重要参与者和见证人，不知不觉已经走过了七年光阴。这七年，全体评委零薪酬参与奖项作品评审，每年三次不定期举行评委会议———月第二个星期六为初审会、四月下旬为终审会和颁奖典礼、夏季不定期总结会，逐年完善严谨的评审规则和公正的评价体系，建立评价标准，制定大奖章程。

我见证评选项目时，评委严肃认真对待每一个项目，会因为观点不同而争论得面红耳赤，会因为达成一致见解而不由自主地站起来一起鼓掌；制定流程、标准、章程等文件时，他们字斟句酌，甚至会严格对待文件格式的标准，事无巨细、严苛认真地对待每一个细节。

奖项从创建到发展至今，重要且坚定的支持者邢同和老师在奖项创立之初，被全体评委推举为主席。几年来，邢同和老师尽心竭力亲自审阅每一份申报作品，参与所有评委相关会议并给予客观且高瞻远瞩的建议；评选第五届奖项时，他更是积极推进了大胆创新的组织变革建议，那一年，评委团不记名投票产生了九名执行委员和三名联席主席。此书的出版，是全体评委对邢同和老师致以的最高敬意。历届评委名单，请见附录。

借此书出版之际，请允许我代表全体评委和地建师小伙伴们感谢所有提交参赛作品的海内外机构，感谢各企业负责人的大力支持，感谢参赛团队伙伴日以继夜地提交奖项申报资料以及工作过程中的积极配合。

在「CREDAWARD 地产设计大奖·中国」的成长道路上，更要感谢在奖项发展前期曾无条件给予财力和活动现场支持的上海博华国际展览有限公司的领导层：王明亮、鲁孟雄、范海燕、杜海燕和他们的伙伴们。

最后，感谢多年来一路风雨兼程走来的团队伙伴们，他们一直拼搏努力。我可以坦言，我们对得起自己的责任和使命。

请允许我把他们的名字一一列上，他们一直在幕后，首次登上"舞台"：龚晓雯、卢菲菲、姬莞婷、王晓君、邵丽、滕雯音、曹雨薇、金童、吴珊、张健、Mimmo Girolamo Barbaccia。

钟 琴

PREFACE II

Qin Zhong

Time flies, it has been seven years since I became an important participant and witness of *CRED*AWARD. During the past seven years, all jurys, as the volunteers, have been involved in the full jury meeting for the review of reward works. The full jury meeting is held three times a year irregularly: The second Saturday of January of each year has the preliminary review meeting, the end of April of each year has the final review meeting and the award ceremony, and each summer has an irregular summary meeting.

I have seen through *CRED*AWARD emerged from nothing: All jurys jointly made the evaluation criteria, the selection process and mechanism, and established its charter. I also saw: When selecting the projects, the jurys seriously reviewed each project, intensely debating due to the different opinions and rising to applaud after reaching the consensus; when setting the procedure, criteria, chapter and other files, they carefully chose words and formats.

Professor Tonghe Xing was elected as the chairperson by all the jurys at the very beginning of *CRED*AWARD. Over those years, Mr. Xing personally reviewed every submitted work, participated in all relevant meetings and gave objective and far-sighted suggestions. For example, at the fifth session, he actively proposed a positive suggestion: That year, the jury produced nine executive committees and three co-chairs. Based on the above facts, all the executive committee members proposed that this book shall pay honor to Mr.Tonghe Xing. Please see the appendix for the list of past jurys.

On behalf of all the jurys and architects, I would also like to express my thanks to all the institutions at home and abroad for submitting their works to us, the heads of the companies for their strong support, and the team members for their active cooperation and submitting award application materials around the clock.

In the growing path of *CRED*AWARD, I would also like to thank Mingliang Wang, Mengxiong Lu, Haiyan Fan, Haiyan Du and so on, the leaders from Shanghai Sinoexpo Informa Markets International Exhibition Co., Ltd., for their unconditional support with finance and the venue support in the first phase of *CRED*AWARD.

Finally, I would like to send my thanks to our team members who supported each other to go through all the ups and downs in the past seven years for their hard work. Now I would like to speak out our hearts that we have realized our responsibilities and missions.

Please allow me to list their names: Xiaowen Gong, Feifei Lu, Wanting Ji, Xiaojun Wang, Li Shao, Wenyin Teng, Yuwei Cao, Tong Jin, Shan Wu, Jian Zhang, Mimmo Girolamo Barbaccia.

学者代表寄语
VOICE OF ACADEMIC REPRESENTATIVES

Jianguo Wang
王建国

大规模的城市化发展，给开发商和建筑师更多机会参与，希望[CREDAWARD 地产设计大奖·中国]坚持以建筑学审美兼顾经济效益的评价标准，传播优秀项目前沿设计理念，树立地产设计行业正确价值取向。

The large-scale urbanization development has given developers and architects more opportunities to participate. I hope that CREDAWARD adheres to the evaluation criteria of architectural aesthetics and economic benefits, disseminates the cutting-edge design concept of outstanding projects, and establishes the correct value orientation of the real estate design industry.

Jianmin Meng
孟建民

希望[CREDAWARD 地产设计大奖·中国]成为倡导"本原设计"的重要平台，将"健康""高效""人文"要素作为评价参评作品的准绳与标尺。

I hope that CREDAWARD, the China Real Estate & Design Award, will become an important platform for promoting "original design", taking the elements of "health", "efficiency", and "humanities" as the criteria and scale for judging works.

Weimin Zhuang
庄惟敏

希望[CREDAWARD 地产设计大奖·中国]，不仅能汇集地产设计理念，更能引领地产设计思想。

I hope that CREDAWARD, the China Real Estate & Design Award, will not only bring together real estate design concepts, but also lead real estate design ideas.

|《中国地产设计集萃 I（2014-2019）》
BEST WORKS OF CREDAWARD I (2014-2019)

目 录　Contents

- 评价标准 —— 014
  EVALUATION CRITERIA

- 奖项历程 —— 016
  JOURNEY TO CREDAWARD

- 项目展示 —— 018
  PROJECT PRESENTATION

  - **01** 公建项目 —— 018
    PUBLIC CONSTRUCTION PROJECT

  - **02** 居住项目 —— 156
    RESIDENTIAL PROJECT

  - **03** 景观设计 —— 206
    LANDSCAPE DESIGN

  - **04** 室内设计 —— 252
    INTERIOR DESIGN

  - **05** 照明设计 —— 292
    LIGHTING DESIGN

302 —— CREDAWARD 地产设计大奖·中国 金银奖年鉴（2014—2019）
GOLD & SILVER AWARDS OF CREDAWARD CATALOGUE(2014-2019)

356 —— 评委名单
JURYS LIST

366 —— 评委寄语
MESSAGE FROM JURYS

372 —— 嘉宾寄语
MESSAGE FROM GUESTS

| 评 价 标 准
EVALUATION CRITERIA

建筑审美
Architectural Aesthetics

环境优先
Environment Priority

评价考量范围参考依据：

· 项目设计具有积极的创新性、原创性、文化传承的体现；

· 项目雅俗共赏，对行业审美趋势具有引领作用；

· 项目功能合理，具有时代性及独特性特征；

· 项目水准对城市及周边环境提升产生积极作用；

· 项目对行业可持续发展起到引领作用；

· 项目在当下甚至将来都具有一定的学习典范作用；

· 项目设计品质与成本兼顾，对产品价值提升产生积极影响。

◉

客 户 体 验

User Experience

◉

经 济 效 益

Economic Benefits

Reference for Evaluation :

· The project should be innovative, original and cultural;
· The project should be both elegant and popular, and play a trendy role in the industry;
· The project should have reasonable functions, featured with the characteristics of the times and uniqueness;
· The project should have a positive effect on the improvement of the city and its surroundings;
· The project should play a leading role in the sustainable development of the industry;
· The project should be a model for others to learn at present and even in the future;
· The project should consider both quality and costs, and have a positive impact on the enhancement of product value.

奖项历程
JOURNEY TO *CRED*AWARD

奖项申报类别修正：修正为公建项目、居住项目、景观设计。

Revised the Category of *CRED*AWARD Application: The revised version covers public construction project, residential project, and landscape design.

参赛企业 86 家　　86 Participating Companies
参赛作品 211 份　　211 Projects
15 大金奖作品　　15 Gold Awards
62 个优秀奖作品　　62 Merit Awards

上海浦东嘉里中心酒店　　参会人数：293 人
Kerry Hotel Pudong Shanghai　　Attendees: 293
The 2nd *CRED*AWARD

第二届地产设计大奖・中国

第一届地产设计大奖・中国
The 1st *CRED*AWARD
上海国际会议中心　　参会人数：149 人
Shanghai International Convention Center　　Attendees: 149

确定评委机制：评委团邀请和退出机制的制定。

Set the Jury Mechanism: the establishment of the rules of the invitation and withdrawal to the jury.

参赛企业 52 家　　52 Participating Companies
参赛作品 285 份　　285 Projects
10 大金奖作品　　10 Gold Awards
68 个优秀奖作品　　68 Merit Awards

第三届地产设计大奖・中国
The 3rd *CRED*AWARD
上海丽思卡尔顿酒店　　参会人数：471 人
The Ritz Carlton Pudong Shanghai　　Attendees: 471

奖项申报类别增加：室内设计；明确参赛作品必须为建成项目。

Added One to the Category of *CRED*AWARD Application: Interior Design. Ensured that the entry must be a completed project.

参赛企业 137 家　　137 Participating Companies
参赛作品 389 份　　389 Projects
20 大金奖作品　　20 Gold Awards
135 个优秀奖作品　　135 Merit Awards

奖项申报类别增加：照明设计。

Added One More to the Category of CREDAWARD Application: Lighting Design.

参赛企业 214 家　214 Participating Companies
参赛作品 658 份　658 Projects
20 大金奖作品　20 Gold Awards
20 大银奖作品　20 Silver Awards
252 个优秀奖作品　252 Merit Awards

上海浦东嘉里中心酒店　参会人数：583 人
Kerry Hotel Pudong Shanghai　Attendees: 583
The 4th CREDAWARD

第四届地产设计大奖·中国

FUTURE···

第五届地产设计大奖·中国
The 5th CREDAWARD
上海浦东嘉里中心酒店　参会人数：1 056 人
Kerry Hotel Pudong Shanghai　Attendees: 1 056

参赛机构 323 家
参赛作品 1 117 份
20 大金奖作品
20 大银奖作品
385 个优秀奖
323 Participating Companies
1 117 Projects
20 Gold Awards
20 Silver Awards
385 Merit Awards

奖项评委团重大改革：
· 成立奖项执委会，产生三名联席主席；
· 邢同和老师担任荣誉主席；
· 建立《奖项章程》，制定更严格科学的评价流程；
· 奖项评价标准文字修订和解读。

Major Reform of the CREDAWARD Jury:
· The CREDAWARD executive committee was established, and three co-chairs were elected.
· In addition, professor Tonghe Xing served as the honorary chairman.
· Established the "CREDAWARD Charter" and set up a more rigorous and scientific evaluation process.
· Revised the words of the CREDAWARD evaluation criteria and interpreted it.

| 项 目 展 示
PROJECT PRESENTATION

01 公 建 项 目 PUBLIC CONSTRUCTION PROJECT

| 01 | 上海环贸广场 IAPM
IAPM SHANGHAI | 020 |
| 02 | 佛山岭南天地商业街区
FOSHAN LINGNAN TIANDI | 028 |
| 03 | 南昌绿地紫峰大厦
NANCHANG GREENLAND ZIFENG TOWER | 040 |
| 04 | 虹桥天地
THE HUB | 048 |
| 05 | 外滩 SOHO
BUND SOHO | 056 |
| 06 | 杭州西子湖四季酒店
FOUR SEASONS HOTEL HANGZHOU AT WESTLAKE | 066 |
| 07 | 北京绿地中心
BEIJING GREENLAND CENTER | 072 |
| 08 | 天环广场
PARC CENTRAL | 080 |

090	保利国际广场 POLY INTERNATIONAL PLAZA	09
098	丹寨万达小镇 DANZHAI WANDA TOWN	10
106	上海徐汇绿地缤纷城 SHANGHAI XUHUI GREENLAND BEING FUN CENTER	11
114	上海外滩金融中心 SHANGHAI BUND FINANCE CENTER	12
120	平安金融中心 PING AN FINANCE CENTER	13
130	周大福金融中心 CTF FINANCE CENTER	14
138	船厂 1862 MIFA 1862	15
146	腾讯总部 TENCENT HEADQUARTERS	16

上海环贸广场 IAPM
IAPM SHANGHAI

评委点评　JURY COMMENTS

这是一个在上海曾经的法租界里开发的大体量的办公商业及公寓综合体。其购物中心在交通流线和体量处理上，成功地再造了城市空间与界面；在内部空间规划和功能定位中，展现了纯熟而又多变的手法，使其成为不可替代的网红景点。充满创意的资产运营与管理则为这一项目带来了持续的发展和演变。

Located at the former French Concession in Shanghai, this multi-functional building is made of many shops, business offices and apartments. Its shopping mall has successfully rebuilt the urban space and interface when it comes to traffic flow and mass treatment. In addition, this building is an irreplaceable internet celebrity attraction thanks to its skillful and various internal space planning and functional positioning. On the other hand, the creative asset operation and management makes it sustainable and developing.

第一届地产设计大奖·中国　公建项目 – 金奖
Public Construction Project | Gold Award of the 1st *CRED*AWARD

业主单位：新鸿基地产	Owner: Sun Hung Kai Properties
规划设计：Benoy 贝诺	Planning Design: Benoy
建筑设计：Benoy 贝诺	Architectural Design: Benoy
室内设计：Benoy 贝诺	Interior Design: Benoy
项目地点：中国·上海	Location: Shanghai, China
基地面积：58 628 m²	Site Area: 58,628 m²
建筑面积：约 222 000 m²	Gross Floor Area: About 222,000 m²
建成时间：2012 年	Date of Completion: 2012

环贸广场 IAPM 设有开阔的屋顶露台、绿化景观以及下沉式花园和冬季花园，访客可前往位于地下层的购物空间和地铁中转站。商场和办公楼的公交接驳不仅可大幅度推动环保及可持续发展的绿色出行，更是创新地库零售层的独特机会。本项目利用媒体科技将林荫大道购物体验延伸至地下，在增加业主商业效益的同时便利访客。环贸广场 IAPM 商场为上海的都市建筑和文化氛围锦上添花，且获得 LEED 金级认证。

IAPM offers a large open roof terrace, greenery landscape, a sunken garden and a winter garden for its visitors. Visitors can access to the underground shops and subway transfer station. The buses reach those underground shops and offices to greatly promote green traffic, to create a new way of doing business for retailing as well. In addition, media technology is used for extending the shopping experience of the boulevard to the underground, increasing the owners' business benefits and bringing convenience for the visitors. IAPM is like the icing on the cake for Shanghai's urban architecture and culture, moreover, it has obtained LEED gold certification.

室内设计旨在营造时尚而优雅的风潮,与商场内高端零售氛围相得益彰。在色彩方面保持柔和的中性色,有效地将顾客的视觉焦点转移到商品上。玻璃栏杆扶手和巧妙的布局呈现了商场内部开阔的视野,令访客能轻松便捷地浏览各大商户。

The interior design aims to create a stylish and elegant trend to perfectly match the high-end retail goods in the mall. To have customers' eyes catching goods, the soft neutral color is used. In addition, the glass handrails and the delicate layout make the mall more spacious, letting customers go shopping pleasantly and conveniently.

商场内部多层 LED 墙和一个个凸出的零售餐饮盒子,则与外墙设计相互呼应。

The multi-layer LED walls inside the mall and the highlighted retail catering boxes echo the design of its exterior wall.

佛山岭南天地商业街区
FOSHAN LINGNAN TIANDI

> **评委点评　JURY COMMENTS**
>
> 岭南天地的设计大胆而不失细腻，将保护建筑、历史建筑与新建建筑和谐融为一体的同时活化了岭南文化。岭南天地创造了多处尺度亲切宜人的公共空间，成为了该项目的焦点，并为佛山的城市生活作出了突出贡献。岭南天地是为数不多的在建筑、文化参与、敏感性设计、场所营造、经济发展、可持续性发展和创造富有活力的生活方式等多个维度均取得成功的优秀案例。
>
> The design of the project, Foshan Lingnan Tiandi (LNTD), is bold and delicate, because it harmoniously integrates the protected buildings, the historical buildings into the newly-built buildings, and activates Lingnan culture as well. It is the massive friendly and pleasant public spaces created by this project that become the spotlight of this project, making tremendous contributions to the urban life of Foshan. Moreover, the project, Foshan Lingnan Tiandi, is one of a few of outstanding cases that have succeeded in multiple aspects such as architecture, cultural participation, sensitive design, place creation, economic development, sustainable development and the creation of a vibrant lifestyle.

第二届地产设计大奖·中国　公建项目 – 金奖
Public Construction Project | Gold Award of the 2nd CREDAWARD

业主单位：瑞安房地产	Owner: Shui On Land
规划设计：Skidmore，Owings & Merrill（SOM）	Planning Design: Skidmore, Owings & Merrill (SOM)
建筑设计：伍德佳帕塔设计咨询（上海）有限公司　巴马丹拿集团	Architectural Design: Ben Wood Studio Shanghai　P&T Design and Engineering Limited
景观设计：地茂景观设计咨询（上海）有限公司	Landscape Design: Design Land Collaborative Ltd.
照明设计：伍德佳帕塔设计咨询（上海）有限公司	Lighting Design: Ben Wood Studio Shanghai
项目地点：中国·广东·佛山	Location: Foshan, Guangdong, China
项目规模：55 020 m²	Project Scale: 55,020 m²
建筑面积：61 219 m²	Floor Area: 61,219 m²
景观面积：4 000 m²	Landscape Area: 4,000 m²
建成时间：2013 年	Date of Completion: 2013
设计周期：36 个月	Cycle of Design: 36 months

岭南天地是位于中国广东省佛山市禅城区的综合性城市再开发项目，为满足采光和通风的要求对建筑区域进行翻新或改建，引入现代化的便利设施传承岭南文化。通过保留佛山传统的建筑特色、手工艺、城市肌理和特色传统活动，岭南天地延续历史街巷的风貌，切实承担起传扬岭南文化的重任。作为佛山地标项目的岭南天地，以其高品质的商业结构和丰富的文化元素有效实现本土与国际、传统与现代、古建筑与多媒体的完美融合，在青砖古巷之间焕发现代艺术的魅力，为佛山精心打造了一个优质的市中心综合发展项目。

Located in Chancheng District, Foshan City, Guangdong Province, China, LNTD is a comprehensive urban redevelopment project. To meet the requirements for lighting and ventilation, LNTD was rebuilt based on introducing the modern facilities and inheriting Lingnan culture. The project continues the style of the historic streets and lanes by preserving the traditional architectural features, handicrafts, urban texture, and traditional activities in Foshan, sincerely undertaking the important task of spreading Lingnan culture. As a landmark project in Foshan, LNTD, with its high-quality commercial structure and rich cultural elements, effectively realized the perfect integration of the local and international, the traditional and modern, ancient architecture and multimedia, reflecting the charm of contemporary arts from the ancient brick alleys, making it a high-quality comprehensive development project in the center of Foshan city.

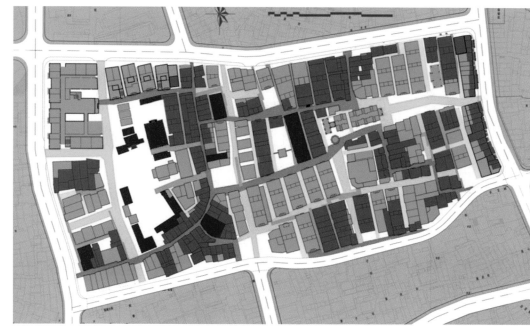

| 主要街巷 Main Streets and Alleys | 小巷 Small Alleys | 新增大巷 Added Big Alleys | 新增小巷 Added Small Alleys | 保护建筑 Preserved Buildings | BWSS 别的保护建筑 BWSS Other Preserved Buildings | 传统风格建筑再现 Restored Urban Typology |

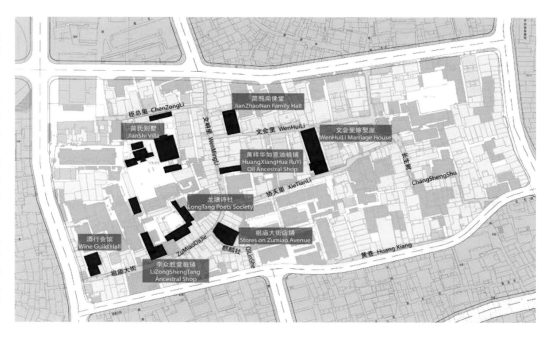

瑞安房地产力争在岭南天地项目中实现四大目标：项目成功商业化、促进历史名城佛山的物质和经济复兴、传承岭南文化、符合佛山核心城区可持续发展的总体规划。

为了实现目标，瑞安房地产遵循了九大总体规划原则：社区规模化的发展、混合用途社区、宜步行的街道和街区、良好的公共交通连接、令人难忘的公共空间和场所、打造地标、尊重当地文化、可持续发展和分阶段开发。

Shui On Land seeks to fulfill four goals for LNTD: to make the project commercially successful, to facilitate the economic growth of the Historic City of Foshan, to inherit Lingnan culture, and to create a master plan for sustainable development of the core area of Foshan.

To reach these goals, Shui On Land follows the nine principles of its master planning: developing model communities, multi-functional communities, pedestrian-friendly streets and blocks, excellent public transit system, memorable public spaces and places, creating landmarks, respecting local culture, sustainable development, and phased development.

岭南天地商业街区的设计运用多种现代化的手法保育和改造片区内 22 幢国家级、省级及市级文物建筑及 128 座优秀历史建筑，在保留大量文物保护建筑和历史建筑的基础上，梳理原有街巷肌理，有机地加入新建筑、新元素。

项目大量保留、再利用原有建筑，再利用旧建筑材料；大量保护保留原有树木，种植本地植物；使用太阳能灯具，进行中水收集并用于植物灌溉和地面清洗。另外项目独创了蒸发冷却风扇系统，在低能耗的前提下有效降低室外温度，以在炎热的天气鼓励人们走出户外，享受自然。项目获得 LEED CS 金级认证。

The design of LNTD uses a variety of modern techniques to preserve and renovate 22 national, provincial and municipal cultural relic buildings and 128 outstanding historical buildings in this area. Based on preserving a large number of the protected cultural relic buildings and historical ones, considering the original texture of streets and lanes as well, the new buildings and elements are organically added to LNTD.

The project largely preserves and reuses the original buildings, recycles the old building materials, protects and retains tremendous original trees, plants the native plants, uses solar lamps, and collects reclaimed water for plant irrigation and ground cleaning. In addition, the project has created an original evaporative cooling fan system, which effectively reduces the outdoor temperature with low energy consumption, to encourage people to go out and enjoy nature in hot weather. The project obtained the gold LEED CS certification.

岭南天地商业街区典型立面图　　Typical elevation of LNTD

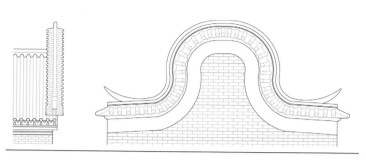

镬耳山墙
Guo-Er gable wall

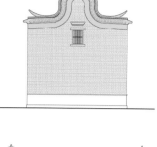

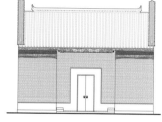

不同的地域文化、气候特征，造就了岭南传统建筑语汇特有的DNA。而传统DNA的保留，成就了项目的独特性，包括镬耳山墙、灰塑、砖雕、樘龙门等。新建筑也大量使用本地材料。

The special DNA of the language of Lingnan traditional architecture is sourced from various local cultures and climatic features. It includes the Guo-Er gable wall, Hui-su, brick caving, Tang-Long Door, etc. Most materials are also local sourced and applied into new buildings.

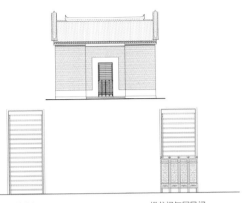

樘龙门
Tang-Long Door

樘龙门与屏风门
Tang-Long Door & Outside Waist Door

墙楣
Qiang-Mei / Wall Brow

石材勒脚
Stone Baseboard

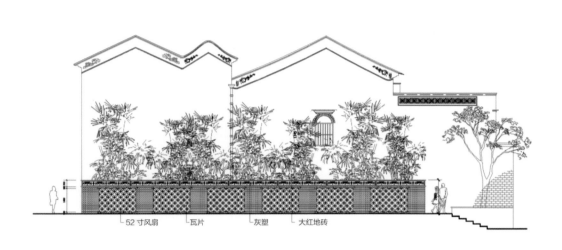

海拔   elevation

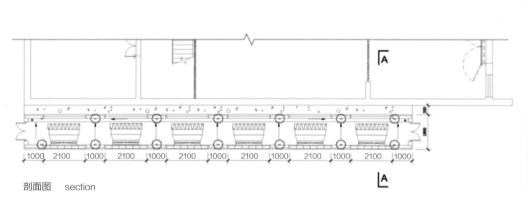

剖面图   section

岭南天地切实承担起传扬岭南文化的重任,传统商户的选择对文化传承有着积极影响。一百多年前,岭南天地所在地作为华南地区的经济文化中心,处于鼎盛时期,这些传统商户繁荣发展,扩张到海外市场并逐渐退出国内市场。瑞安房产在吸引传统商户回归方面付出了巨大的努力,陆续在新加坡和其他东南亚等国家的华裔中找到其后人,并将这些传统品牌带回岭南天地。

Shui On Land made extensive efforts to curate many of the original businesses that once flourished over a hundred years ago during the heydays of LNTD as the cultural and economic center of Southern China but had disappeared in China. Shui On Land located many such businesses in ethnic Chinese in places like Singapore and other South eastern Asia countries and brought them back to LNTD.

南昌绿地紫峰大厦
NANCHANG GREENLAND ZIFENG TOWER

❝ 评委点评   JURY COMMENTS

项目形体简洁优美，功能布局合理，设计采用形体凹进与幕墙表皮肌理对比的设计手法，面向主要城市界面塑造了独特的城市之窗形象，在标志性、独特性塑造上独具特色。

With a simple and beautiful shape and reasonable functional layout, the design uses the contrast method, such as, the concave shape and the surface texture of the curtain wall, presenting a unique city window for the main urban interface, displaying its iconic and unique characteristics. ❞

第二届地产设计大奖·中国　公建项目 – 金奖
Public Construction Project | Gold Award of the 2nd *CREDAWARD*

业主单位：绿地控股集团	Owner: GREENLAND HOLDING GROUP
规划设计：Skidmore, Owings & Merrill（SOM）	Planning Design: Skidmore, Owings & Merrill (SOM)
建筑设计：Skidmore, Owings & Merrill（SOM）	Architectural Design: Skidmore, Owings & Merrill (SOM)
景观设计：SWA	Landscape Design: SWA
室内设计：HBA	Interior Design: HBA
合作设计：华东建筑设计研究总院	Joint Design: ECADI
项目地点：中国·江西·南昌	Location: Nanchang, Jiangxi, China
项目高度：267.98 米	Project Height: 267.98 m
建筑面积：209 058 m²	Floor Area: 209,058 m²
建成时间：2015 年	Date of Completion: 2015
设计周期：12 个月	Cycle of Design: 12 months

南昌绿地紫峰大厦地处南昌市的重要位置,项目地块南至紫阳大道,东至创新一路,为人们到达自然休闲区提供了良好的出入通行系统以及便捷的交通。

项目设计充分考虑与城市及周围环境之间的关系。作为南昌市高新区的标志性塔楼,设计不仅要考虑向外远眺的景色,还要考虑到高新科技区未来的商业核心区的重要作用。配合标志性的建筑造型,南昌绿地紫峰大厦已成为南昌市的一座重要且广受欢迎的标志性建筑。

本项目采用的倾斜玻璃表面烘托出"城市之窗"的独特景象,与建筑外围护巧妙地相呼应。

Nanchang Greenland Zifeng Tower is situated in a critical location in Nanchang City. The project plot extends to Ziyang Avenue in the south and Chuangxin I Road in the east, which provides a good access system and convenient transportation for people to reach the natural recreation area.

The design of this project fully considers the relationship among the building, the city and the surroundings. As a landmark tower in Nanchang, the design not only is sensitive to the distant views from the Central District, but also anchors the potential commercial core of the High-tech Industrial Park. Combined with an iconic architectural shape, Nanchang Greenland Zifeng Tower has become an important and popular destination in Nanchang.

The inclined glass surface used in this project brings out the unique scene of the "city window", and subtly echoes the building envelope.

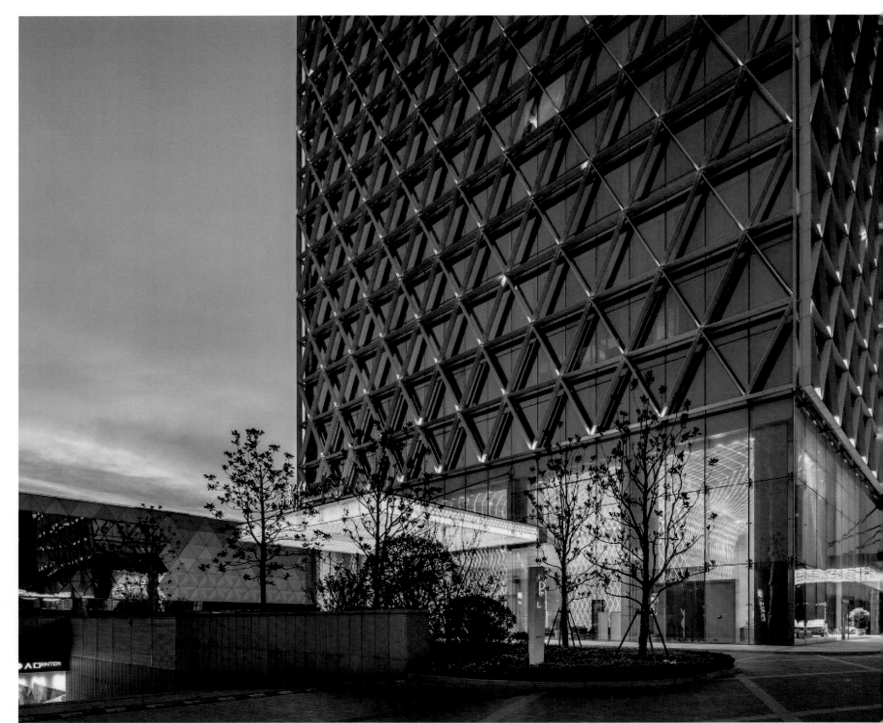

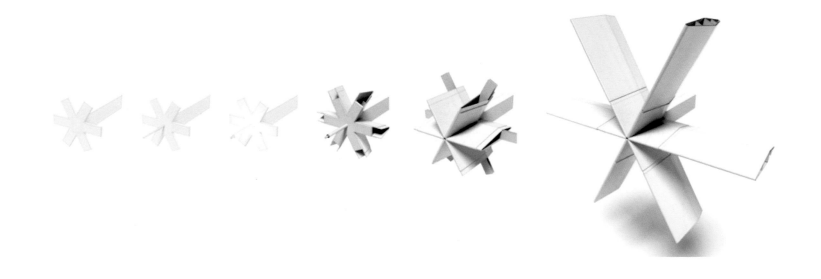

塔楼四面设计了三角形遮阳肋，不仅在建筑立面上形成了独特的肌理，而且也确保了提供一座性能更佳的建筑。横向和斜向遮阳板在很大比例上消除了太阳增热。除了遮阳板外，项目还合并了其他节能设计，如低放射率玻璃、太阳能热水管及太阳能采电板。

Triangular shading ribs are designed on all four sides of the tower to form a unique architectural texture and ensure a better performing building. The use of both horizontal and diagonal fins eliminates a large percentage of the solar heat gain. In conjunction with these shading devices, the project incorporates a variety of other energy saving strategies, such as low-E glass, evacuated tubes and the potential use of photovoltaic panels for energy harvesting.

三角形元素为设计主题,贯穿从总体设计、景观设计、立面幕墙、室内设计到标识设计由外而内的全过程设计,高度保持了设计创意的完整性,给人以一气呵成、行云流水的视觉效果。

Triangular elements reoccur as a design theme throughout the building, including overall design, the landscape design, curtain wall facades, interior design and logo design, making the overall design coherent and graceful .

虹桥天地
THE HUB

> **评委点评　JURY COMMENTS**
>
> 虹桥天地的成功之处在于其充分打通了室内外体验，创造了充满活力的公共空间，这一特质不仅体现在整体的混合功能社区中，也在其室内设计中一以贯之。室外庭院、室内广场、步行连廊以及中庭等多样的空间被各式活动充分激发了活力，有效提供了运营场地。这些室内外公共空间丰富多元而又生机勃勃，使虹桥天地从普通的商场中脱颖而出。
>
> As a mult-functional community, the success of THE HUB is that it combines indoor and outdoor experiences and creates vibrant public places. This characteristic is also clearly manifested in its interior design. The exterior courtyards, interior plazas, pedestrian corridors, and atriums are filled with various activities which energize their functions and provide efficient operations. The diverse and vibrant atmospheres of interior and exterior places make THE HUB stand out from regular retail environments.

第二届地产设计大奖·中国　公建项目 – 金奖
Public Construction Project | Gold Award of the 2nd *CRED*AWARD

业主单位：瑞安房地产	Owner: Shui On Land
规划设计：Ben Wood Studio Shanghai	Planning Design: Ben Wood Studio Shanghai
建筑设计：巴马丹拿集团 　　　　　Ben Wood Studio Shanghai	Architectural Design: P&T Group in collaboration 　　　　　Ben Wood Studio Shanghai
景观设计：地茂景观设计咨询（上海）有限公司	Landscape Design: Design Land Collaborative Ltd.
室内设计：巴马丹拿集团 – 办公楼 　　　　　G+A- 大堂设计 　　　　　CallisonRTKL- 购物中心	Interior Design: P&T Group-Office 　　　　　G+A-Main Lobby 　　　　　CallisonRTKL-Mall
项目地点：中国·上海	Location: Shanghai, China
占地面积：62 000 m²	Site Area: 62,000 m²
建筑面积：380 000 m²	Floor Area: 380,000 m²
景观面积：10 238 m²	Landscape Area: 10,238 m²
建成时间：2015 年	Date of Completion: 2015
设计时间：60 个月	Cycle of Design: 60 months

虹桥天地位于虹桥商务区核心位置,项目规划总建筑面积约 380 000 平方米,规划设计集展示办公、购物、餐饮、休闲、娱乐、演艺于一体的一站式新生活中心,为虹桥商务区的办公人群、西上海居民以及高铁一小时经济圈所辐射的 7 500 万人口提供商务、休闲、娱乐的新型地标平台。虹桥天地更注重可持续发展,还在地面与地下建造了约 39 500 平方米的公共空间,成功打造了一个面向多元社群的活力社区。

Located in the core of the Hongqiao Central Business District, Shanghai, with 380,000 m^2 construction area, THE HUB, as the one-stop new life center for offices, shopping, dining, leisure, entertainment, and performing arts, provides a new landmark platform for the office staff who work at the Hongqiao Central Business District, the residents who live in the western Shanghai, and the 75 million people who reach THE HUB by the high-speed train within one hour to do business and enjoy leisure and entertainment. At the same time, THE HUB pays more attention to its sustainable development. It has also built about 39,500 m^2 public space covering the ground and the underground, successfully creating a vibrant community for diverse communities.

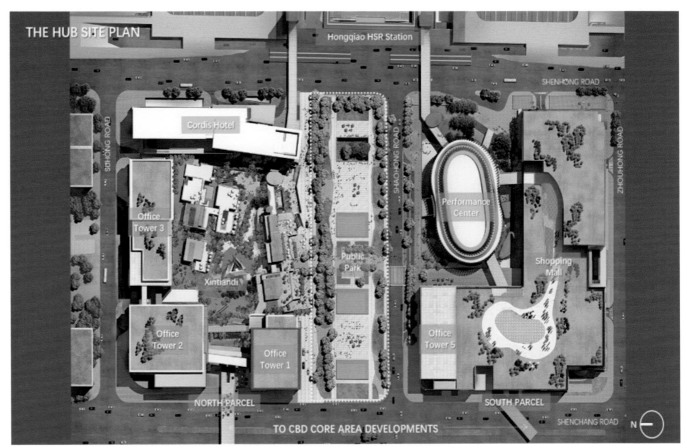

· 以"连接"为主旨，真正功能复合的 TOD 项目
虹桥天地是首个与虹桥交通枢纽直接相连的商业综合体。项目整体设计以"连接"为主旨，采用地下、地上多种方式与虹桥火车站便捷相连，实现了商场、街区商业、展示办公、酒店、文化演艺在水平和垂直方向上的复合利用，是一个真正功能复合型的 TOD 项目。

· The Multi-Functional Tod Project With The Theme Of Linking
As the first commercial complex area directly connected to the Hongqiao Transportation Hub, THE HUB, designed with the theme of "linking", has the underground and ground access to the Hongqiao Railway Station, horizontally and vertically offering many business places, such as shopping malls, commerce blocks, showroom offices, hotels, and cultural performances. It is a truly multi-functional TOD project.

外滩 SOHO
BUND SOHO

评委点评   JURY COMMENTS

作为百年外滩的收官之作，外滩 SOHO 的开发与建设承载了历史的责任。建筑师采用了现代的设计手法，通过空间序列的组合，在外滩创造了一个富有历史神韵的当代地标建筑。这一项目对于各大城市中正在进行的城市更新提供了重要的借鉴。

As the final work of more than one hundred years Bund in Shanghai, the development and construction of the SOHO on the Bund bear historical responsibility. With the modern design techniques, the architect creates a contemporary landmark building full of historical charm through the combination of spatial sequences, providing an important example for the ongoing urban renewals in big cities.

第二届地产设计大奖 · 中国   公建项目 – 金奖
Public Construction Project | Gold Award of the 2nd CREDAWARD

业主单位：SOHO 中国	Owner: SOHO China
建筑设计：gmp · 冯 · 格康，玛格及合伙人建筑师事务所	Architectural Design: gmp · von Gerkan, Marg and Partners Architects
联合建筑设计：华东建筑设计研究院总院	Co Architectural Design: ECADI
联合景观设计：德国雷瓦德景观建筑事务所	Co Landscape Design: Rehwaldt Landschaftsarchitekten
景观设计：澳洲艺普得城市设计咨询有限公司	Landscape Design: INTEGRATED PLANNING AND DESIGN PTY LTD
照明设计：德国 Conceptlicht 照明设计公司	Lighting Design: Conceptlicht GmbH
幕墙设计：莫特麦克唐纳有限公司	Facade: Mott Macdonald
摄　　影：Christian Gahl	Photographer: Christian Gahl
项目地点：中国 · 上海	Location: Shanghai, China
建筑面积：189 509 m²	Floor Area: 189,509 m²
景观面积：3 944 m²	Landscape Area: 3,944 m²
建成时间：2015 年	Date of Completion: 2015
设计时间：60 个月	Cycle of Design: 60 months

项目位于上海最著名的城市大道"外滩"南端,毗邻历史保护街区。外滩 SOHO 是一个涵盖了办公、商业、娱乐的城市综合体项目。四栋高度在 60 米到 135 米之间的高层办公建筑形成了略长的富有变化的项目体量,与两栋商业裙房营造出一个带有小型广场、可通往外滩江边的生动城市空间。

The site is located at the southern end of the Bund which is a famous boulevard in Shanghai and adjacent to a preserved historic block. The project of Bund SOHO is a mixed-use urban developing one, covering offices, retail and entertainment. The four high-rise office buildings with heights between 60m and 135m take shape in longish, shifted volumes, creating a vivid urban space with small squares and alleys leading to the waterfront of the Bund.

外滩 SOHO 办公和商业综合体作为上海黄浦江畔外滩南段最后一座建筑，以优雅的剪影收束了被赋予上海"东方巴黎"盛誉的外滩天际线，和谐地融入了外滩建筑群从新哥特主义到"装饰艺术"风格的多元风格中。建筑的设计理念在于延续万国建筑群的历史风格，但避免刻意的怀旧和现有形式的重复，并在以古城公园和豫园为标志的老城区之前定义出历史建筑群的尾声。在上海近年建成的大量单体建筑中，外滩 SOHO 为守护传承城市地域文脉作出了贡献。

As the final building at the southern Bund along the Huangpu River in Shanghai, the SOHO on the Bund (or SOHOB, for short), together with its plaza, encloses the Bund Skyline known as the "Oriental Paris" with its elegant silhouette, melting in the buildings on the Bund featured with diverse styles from Neo-Gothic to Art Deco. Its design concept is to inherit the historical style of the old buildings along the Bund, but avoid deliberate nostalgia and repetition of existing forms, and mark the ending of the historical buildings ahead of the old city labelled by Gucheng Park and Yu Garden. It is obvious that SOHOB plays an important role to protect and inherit the local culture among the large number of mono-buildings built in Shanghai in recent years.

立面顶部玻璃后退，凸显石柱的竖向效果。

The top glass of the facade is set back to emphasize the vertical effect of the stone columns.

· 新老外滩过渡——传统延续与现代转译

外滩 SOHO 作为城市地标，标志了新老外滩的过渡。设计以对周边历史环境例如典型的条形"弄堂"石库门建筑、通往江边的狭窄的街道网、经典的天际线、老建筑的竖向石材立面等深入研究为指导，将老外滩典型历史建筑立面中的竖向构图与三段式布局进行现代转译，用米色天然石延续了外滩的传统质感。

· Transition From The Old Bund To The New One——Inherting Tradition And Reinterpreting In Modern Way

As an urban landmark, SOHOB marks the transition from the old Bund to the new one. Its design ideas are shown below: After having researched its historical surroundings covering the Shikumen building which is a typical bar-shaped "alley", the narrow street network leading to the river, the classic Bund Skyline, and the vertical stone facade of the old buildings, the design, from modern view, had reinterpreted the vertical composition and three-stage layout of the facade of the typical historical buildings of the old Bund, using the beige natural stone to extend the traditional texture of the Bund.

公共空间室内设计延续了外部的建筑风格。

Interior design of public space continues the exterior architectural style.

— 065 —

杭州西子湖四季酒店
FOUR SEASONS HOTEL HANGZHOU AT WESTLAKE

> **评委点评　JURY COMMENTS**
>
> 酒店平面布局疏密有致，结合湖景形成丰富多变的空间层次。建筑采用传统风格现代表达，理性而克制、含蓄不失大气，是新一代园林酒店的典范之作。
>
> Combined with the lake view, the hotel provides a rich and changeable space for its guests thanks to a scientific layout. The hotel is a model of garden hotels in a new generation because it adopts a humble but elegant method to modernize the traditional style.

第三届地产设计大奖·中国　公建项目 – 金奖
Public Construction Project | Gold Award of the 3rd CREDAWARD

业主单位：杭州金沙港旅游文化村有限公司	Owner: Hangzhou Jinsha Harbour Tourism and Cultural Village Co., Ltd.
规划设计：goa 大象设计	Planning Design: GOA
建筑设计：goa 大象设计	Architectural Design: GOA
景观设计：美国 BENSLEY 景观设计公司	Landscape Design: BENSLEY
北京奇思哲景观设计工作室有限公司	HCZ Landscape Architecture P.C.
室内设计：美国 BLD 室内设计公司	Interior Design: BLD
日本 SPIN 室内设计公司	SPIN
照明设计：新加坡 PLD 灯光设计公司	Lighting Design: PLD
摄　　影：郎水龙，泠城摄影工作室	Photographer: LANG Shuilong, shiromio studio
项目地点：中国·杭州	Location: Hangzhou, China
建筑面积：43 537 m²	Floor Area: 43,537 m²
建成时间：2010 年	Date of Completion: 2010
设计周期：72 个月	Cycle of Design: 72 months

杭州西子湖四季酒店环抱"西进"之西湖。近处，与西湖十景之"曲苑风荷"毗邻，水泽氤氲；远处，隔湖与南高峰诸山相望，有"悠然见南山"之意。围绕景观主轴，客房部分沿主体两翼展开，庭院穿插其间，空间节奏变幻。餐饮宴会区在东侧沿内湖一路铺展，中餐厅、包厢等结合游廊布置，移步异景。园中山水以原有框架为基础作调整，叠山引水，重新植栽花木。别墅区相对独立，分两组街坊式布置，街巷空间层次深远。在构造上，木作、瓦作、石作部分基本沿用传统苏式做法，其他装饰细部则重新设计，手法节制，形制简洁，但求意到。

Four Seasons Hotel Hangzhou At West Lake is surrounded by the "westward" West Lake. From the near, it is adjacent to the "Qu Yuan Feng He", which is literally translated into "The Lotus in the Breeze at Crooked Courtyard", filled with water and mist, and being one of the ten scenic spots of West Lake; from the far, it faces the mountains of the South Peak across the lake. Based on the main axis of the landscape, the guest rooms, interspersed by the courtyards, are laid out along the two sides of the main body. The catering and banquet area is arranged along the inner West Lake on the east side, focusing on choosing the attractive spots to set Chinese restaurants, private dining rooms. The rebuilt garden, based on the original frame of the landscape, builds man-made mountains and ponds and replants flowers and trees. When it comes to its structure, the woodwork, tile work and stone work basically follow the Chinese traditional Su Style, while other decorative details are redesigned with delicate, conservative techniques and concise styles to depict the artistic conception.

· **精心设计的景观主轴线**
项目采用传统江南园林风格,将"入口庭院—大堂—泳池—跌水—湖面—别墅区客房—远山"作为酒店的景观主轴。

· **简洁形制描绘意境**
在构造上,木作、瓦作、石作部分基本沿用传统苏式做法,其他装饰细部则重新设计,手法节制、形制简洁,描绘出整体意境。

· **Well-Designed The Main Axis Of The Landscape**
Making best of the style of the traditional garden on the Yangtze Delta, the main axis of the landscape of the hotel follows the line of the "entrance courtyard-lobby-swimming pool-falling water-lake surface-guest rooms of villas-distant mountains".

· **Concisely Depicting The Artistic Conception**
When it comes to its structure, the wood work, tile work and stone work basically follow the Chinese traditional Su Style, while other decorative details are redesigned with delicate, conservative techniques and concise styles to depict the artistic conception.

· 山水是核心，林园就是空间塑造的目标

　　设计强化西湖"如画"的意境，将发掘景观因素和延续自然意境作为酒店主景和营造重点，精心思考和设计建筑群整体关系以及庭院空间的营造。

· The hills and water are its core, the garden is its aim.

The design highlights the picturesque West Lake to explore its landscape elements and keep its natural beauty, carefully designing the overall relationship between the buildings and the courtyard.

北京绿地中心
BEIJING GREENLAND CENTER

> 评委点评 JURY COMMENTS
>
> 项目形体简洁挺拔，功能布局合理，幕墙表皮设计采用中国结的设计理念与城市文脉呼应，通过梯形幕墙板块重复模式形成了独特的表面肌理，以独特的气质成为城市地标典范。
>
> The shape of the project is simple and straight, and its functional layout is reasonable. In addition, the Chinese knots design used for its curtain wall surface echoes the urban context and the repeating pattern of trapezoidal curtain wall panels forms a unique surface texture, making it a model of urban landmarks.

第三届地产设计大奖·中国　公建项目 – 金奖
Public Construction Project | Gold Award of the 3rd CREDAWARD

业主单位：绿地控股集团	Owner: GREENLAND HOLDING GROUP
建筑设计：Skidmore, Owings & Merrill（SOM）	Architecture Design: Skidmore, Owings & Merrill（SOM）
结构设计：Skidmore, Owings & Merrill（SOM）	Structure Design: Skidmore, Owings & Merrill（SOM）
土木工程：Skidmore, Owings & Merrill（SOM）	Civil Engineering Design: Skidmore, Owings & Merrill（SOM）
设备工程设计：Skidmore, Owings & Merrill（SOM）	Equipment Engineering Design: Skidmore, Owings & Merrill（SOM）
景观设计：SWA	Landscape Design: SWA
室内设计：Skidmore, Owings & Merrill（SOM）（主要公共空间部分）	Interior Design: Skidmore, Owings & Merrill (SOM) (only for main public spaces)
照明设计：KGM	Lighting Design: KGM
合作设计：北京六建集团有限责任公司	Joint Design: Beijing Liujian Construction Group
项目地点：中国·北京	Location: Beijing, China
建筑高度：260 m	Architectural Height: 260 m
建筑面积：172 734 m²	Floor Area: 172,734 m²
景观面积：3 944 m²	Landscape Area: 3,944 m²
建成时间：2016 年	Date of Completion: 2016
设计时间：16 个月	Cycle of Design: 16 months

北京绿地中心坐落在北京大望京商业区，是一座具有高度可持续性的综合型地标大厦。建筑高260米，总共55层，着重于可持续性和高效节能性。每个面由两个竖向、低辐射保温透明玻璃模板组成，两个梯形板块随着塔楼的高度向上交替变换角度，令塔楼立面产生了与天空和地面周边环境相互交融的效果。设计方案旨在通过可持续性措施达到比标准大楼节电节水30%的目标，其中包括直接数控大楼自动化系统、热回收轮、采暖和冷却系统的变速泵以及利用蒸发冷却的水侧节能器等。作为一个充分考虑环保需求的综合性城市项目，北京绿地中心是一座视觉上极具吸引力、同时又具有高度灵活性和可持续性的建筑。

The tower of Beijing Greenland Center, located in Downageing business district, Beijing, is a multi-functional landmark building with a height of 260 meters and 55 floors, featured with sustainability and energy efficiency. Each facet of the building is made up of two vertical, low-radiative, insulative and see-through glass plates. Thanks to its two step-plates alternately varying the angles with the height of the tower, the facade of the tower perfectly melts in the sky and the surroundings on the ground. In order to achieve the goal of saving electricity and water by 30% compared with the standard building through sustainable measures by using direct numerical control automation system, heat recovery wheel, variable speed pump for heating and cooling system, and water-side economizer using evaporative cooling. As a comprehensive urban project to fully consider the demands for environmental protection, the tower of Beijing Greenland Center has the traits of being visually attractive, highly flexible, and sustainable.

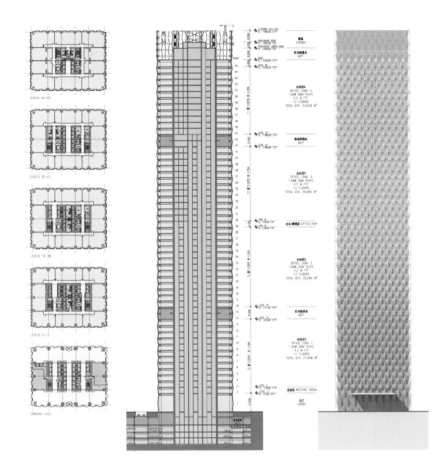

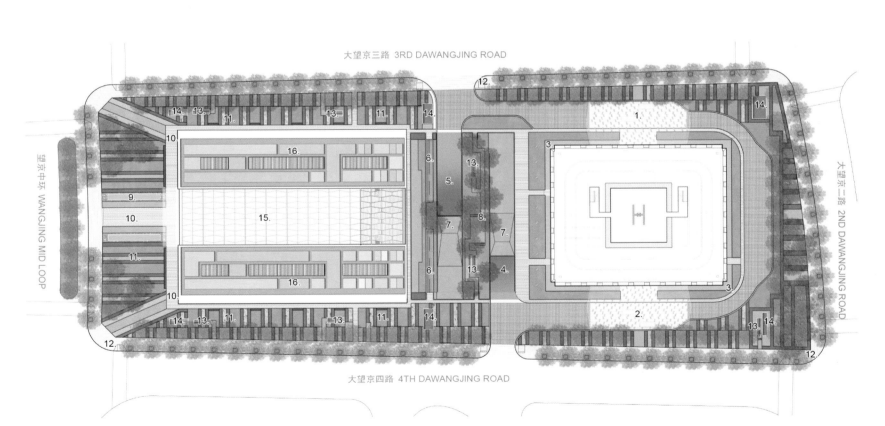

项目拥有可提供自遮阳的菱柱梯形立面造型。设计团队为力求创造一个造型优雅、引人注目的大厦,在简洁造型、采光及遮阳方面进行了许多探索。

Rhombus-step-facade automatically provides shade. To create an elegant and eye-catching building, the design team has carried out many ways to explore its shape, lighting and shading.

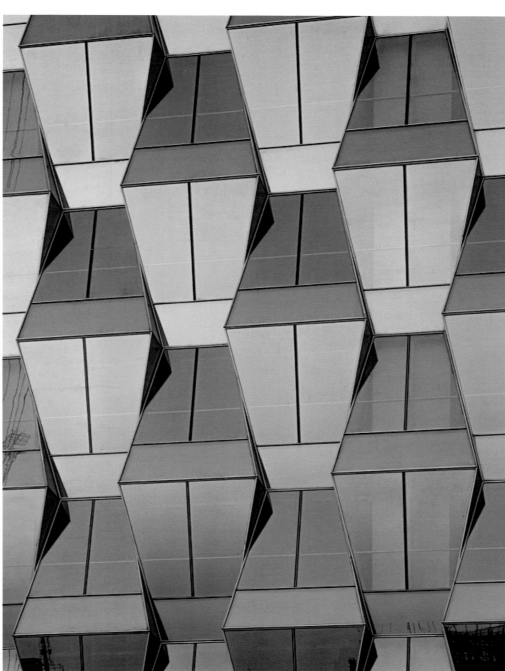

比较分析　　COMPARATIVE ANALYSIS

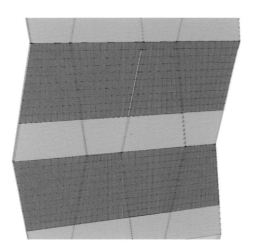
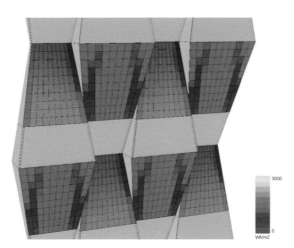

平面立面　FLAT FACADE　　　　　　　　　　起伏阴影立面　UNDULATINGSELF-SHADING FACADE

· 立面捕捉动态光影
生动的大厦立面造型有助于捕捉和反射日间光影，为周围的邻里社区带来勃勃生机。

· 阳光和遮阳的双重要素
项目借用浮雕设计理念，采用等腰梯形模数，起到棱镜的作用，反射和折射光影；通过阳光和遮阳的双重要素，造就了塔楼引人注目的外观效果。

· 立面自遮阳，提高环保能效
由棱柱梯形玻璃单元体组成的大厦立面造型可以提供自遮阳，从而提高了大楼的环保性能。

· Facades Capture Moving Light And Shadow
The vivid facade of the tower helps to capture and reflect the light and shadow of the daytime, bringing vitality to the neighbourhood.

· The Dual Role Of Refraction And Shading
With the help of the relief carving technique, the isosceles-step-modulus is used to serve as a prism to receive and refract light and shadow, playing the dual role of sunlight and shading as well as creating the eye-catching appearance of the tower.

· Automatic Shading To Improve Energy Efficiency
The facade of the tower is made of the rhombus-step-glass-plate , automatically providing shade and improving energy efficiency.

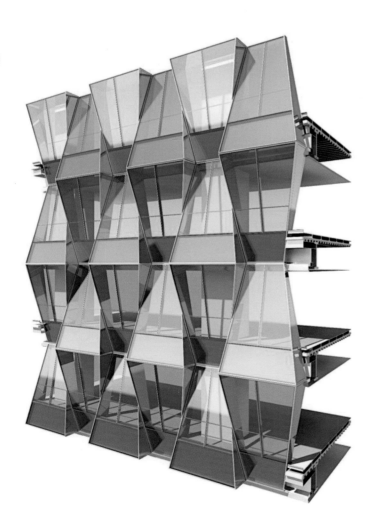

SOUTHEAST
SOUTHWEST

三月
MARCH

东南和西南立面　SOUTHEAST & SOUTHWEST FACADES　　东北和西北立面　NORTHEAST & NORTHWEST FACADES

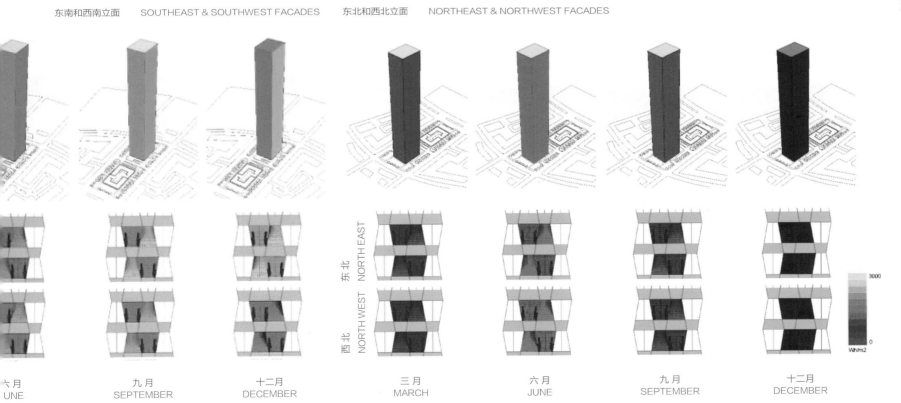

| 六月 | 九月 | 十二月 | 三月 | 六月 | 九月 | 十二月 |
| JUNE | SEPTEMBER | DECEMBER | MARCH | JUNE | SEPTEMBER | DECEMBER |

东北 NORTH EAST
西北 NORTH WEST

天环广场
PARC CENTRAL

> **评委点评　JURY COMMENTS**
>
> 空间连接、景观连接、交通连接、智能连接使得多样化的商业体验和活动有了依托和空间，"双鱼"建筑设计活跃了建筑形态，没有形式主义嫌疑。
>
> It is the delicate design featured with spatial connection, landscape connection, traffic connection, and intelligent connection that provides support and space for diversified commercial experiences and activities. Moreover, the design of the two-fish-shaped buildings enriches architecture forms, naturally displaying the buildings.

第三届地产设计大奖·中国　公建项目 – 金奖
Public Construction Project | Gold Award of the 3rd CREDAWARD

业主单位：新鸿基 & 广州新中轴建设有限公司	Owner: Sun Hung Kai Properties & Guangzhou New Central Axis Construction Co., Ltd.
规划设计：Benoy 贝诺	Planning Design: Benoy
建筑设计：Benoy 贝诺	Architectural Design: Benoy
执行建筑设计：吕元祥建筑事务所	Executive Architectural Design: Ronald Lu & Partners
室内设计：Benoy 贝诺	Interior Design: Benoy
摄　　影：张虔希	Photographer: Terrence Zhang
项目地点：中国·广东·广州	Location: Guangzhou, Guangdong, China
建筑面积：110 424 m²	Floor Area: 110,424 m²
规划用地：47 601.1 m²	Site Area: 47,601.1 m²
绿地面积：17 322 m²	Green Area: 17,322 m²
建成时间：2016 年	Date of Completion: 2016

天环广场高 24 米，是一座低层标志性建筑，建有两层地上空间和三层地下空间。由于项目高度较周边建筑为低，设计上采用抢眼和强烈的视觉效果，为广州新中轴增加韵律感。

建筑设计概念来自中国文化中象征和平、和谐和财富的鲤鱼，不锈钢硬壳结构屋顶酷似两条游动的鱼。双鱼造型大楼环绕中央花园而建，通过人行天桥连接。巨大的树状柱子支撑着屋顶，下方的花园空间将景观元素一直延伸到大楼室内。内设的雨水收集系统、节能低辐射玻璃幕墙和 EFTE 薄膜屋顶也大幅提高了建筑物的环保成效。

With two ground floors and three underground ones, the 24-meter-high PARC Central is a landmark low-rise building. Since the height of this building is lower than that of the surrounding ones, the eye catching, and strong visual effects are adapted to the design to have the new axis of Guangzhou beating.

The architectural design concept comes from the carp symbolizing peace, harmony and wealth in Chinese culture. The double-fish-shaphed building is built around the central garden and is connected by a pedestrian bridge. Huge tree-like pillars support the roof, and the below garden space extends the landscape elements all the way to the interior of the building. The built-in rainwater collection system, energy-saving low-radiation glass curtain wall and EFTE membrane roof also greatly improve the environmental protection effect of the building.

· **标志性双鱼建筑**
本项目设置 40 米的宽度保证视觉走廊的通畅，高度尊重并且加快城市中轴的节拍，形成城市公园与零售功能间的无缝连接，承担起将城市带回低密度发展节奏的重任；利用天环在中心轴线的优越位置和绿化带的建设要求，为城市的商业发展带来一座绿色公园，超过 60% 被绿化覆盖的多层次绿化引领低碳生活，满足城市发展的宏伟愿景；沉浸在都市公园环境下的"购物体育场"与周边交通的关联创造了一种独特的购物体验。

· **Iconic Double-Fish-Shaped Buildings**
The project sets a width of 40 meters ensures the smoothness of the visual corridor, highly respects and speeds up the beat of the city's central axis, forms a seamless mixture between urban parks and retail functions, and assumes the important task of bringing the city back to the low-density development speed. Utilizing the superior location of PARC Central on the central axis and the construction requirements of green belts, a green park is brought to the commercial development of the city. More than 60% of greening leads to low-carbon life and meets the grand vision of urban development. The relationship between the "shopping stadium" immersed in the urban park and the surrounding traffic creates a unique shopping experience.

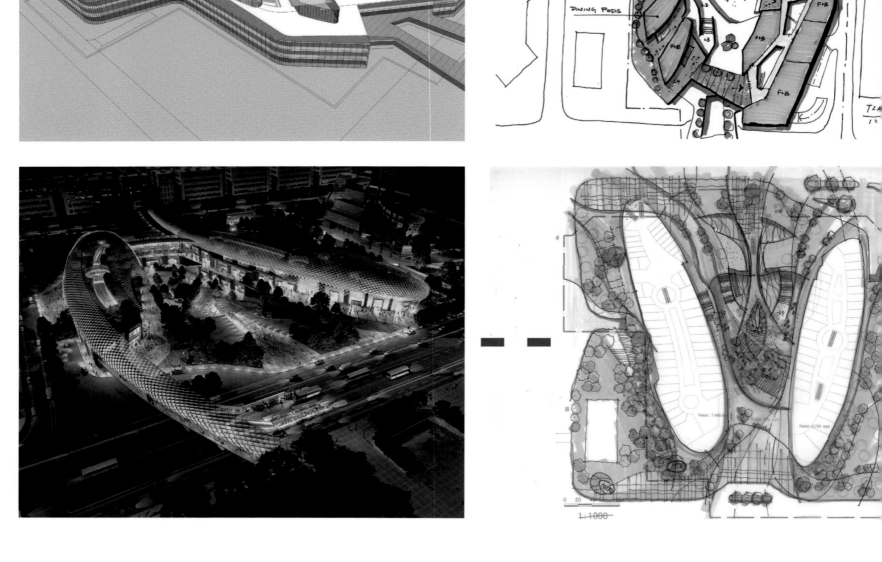

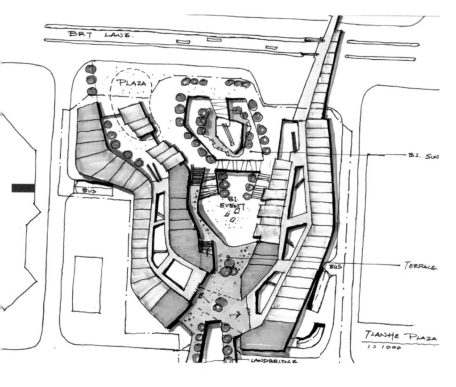

· 设计概念演变

最初的概念是一个 V 形的设计，边缘锋利而连续的建筑体量围绕着下沉式花园。它逐渐演变为具有更柔和边缘并且分离的形式，尊重广州的中央轴线。景观及城市公园元素亦被纳入整个设计演变过程中的核心组成部分。

· Changing Design Idea

To keep and respect Guangzhou's central axis, the first design concept, a V-shaped design with sharp edges and ongoing building surrounding the sunken garden, has been gradually changed into the one with softer edges and separating buildings, taking the landscape and urban park into the core elements of the entire design.

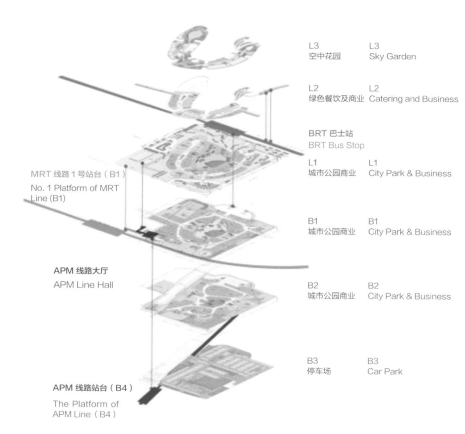

L3 空中花园	L3 Sky Garden
L2 绿色餐饮及商业	L2 Catering and Business
BRT 巴士站	BRT Bus Stop
L1 城市公园商业	L1 City Park & Business
B1 城市公园商业	B1 City Park & Business
B2 城市公园商业	B2 City Park & Business
B3 停车场	B3 Car Park

MRT 线路1号站台（B1）
No. 1 Platform of MRT Line (B1)

APM 线路大厅
APM Line Hall

APM 线路站台（B4）
The Platform of APM Line（B4）

· **公交导向发展项目**

这个以公共交通为主导，总建筑面积110 000平方米的发展项目为广州建立了一个新的公共交通中转站。天环广场与中国地铁系统和公交网络无缝相连，且有架空天桥连接周边的发展项目，交通四通八达。项目的架空、地面及地下通道均无障碍，方便易达，是社交聚会理想之地。

· **Transit-Oriented Developing Project**

The PARC Central, the name of the transit-oriented project, covering an area of 110,000 m², establishes a new public transportation hub for Guangzhou, seamlessly connecting the subway and bus stops as well as the surroundings with its overhead flyovers to make it an ideal place for people gatherings.

· 定制交通连接设计

交通连接是天环广场设计的关键部分，复杂的贝壳风格结构为天环的交通设计带来定制的触感，但对施工是一个巨大的挑战。为了成功实现设计，设计人员对具有特殊弧度的多层夹层玻璃，进行了三次试验，确定了在没有第二层支撑结构的情况下与夹层玻璃直接相连。

· Customized Design For Traffic Link

The key part of the PARC Central design is to connect traffic. In order to make traffic design customized, the complex shell-style structure, which posed a huge challenge to the construction, was used. In addition, to successfully realize the design, the construction took three times to test the technically designed multilayer-laminated-glass with a special arc. On the other hand, the frame had to be designed to directly connect to the laminated glass without a second layer of supporting structure.

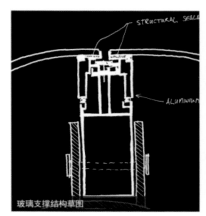
玻璃支撑结构草图
A Drawing of Glass Support Structure

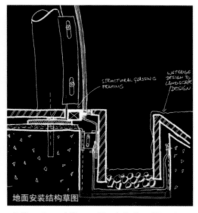
地面安装结构草图
A Drawing of Ground Installation Structure

保利国际广场
POLY INTERNATIONAL PLAZA

> 评委点评　JURY COMMENTS
>
> 项目规划设计考虑了城市肌理和本地块的互动关系，主楼灯笼状建筑与结构设计结合，体现力量美学，创新的可持续性系统积极应对气候和空气质量挑战，成为新办公塔楼的典范。
>
> Not only does the planning and design of the project take into account the interaction between urban texture and this plot, but also presents the beauty of power contributed by the combination of the lantern-shaped main building and its structural design. Moreover, its innovative and sustainable system actively responds to the challenges brought by climate and air quality, making it a real model for new office towers.

第四届地产设计大奖 · 中国　公建项目 – 金奖
Public Construction Project | Gold Award of the 4th *CREDAWARD*

业主单位：保利（北京）房地产开发有限公司	Owner: China Poly Real Estate Company Limited
建筑设计：Skidmore, Owings & Merrill（SOM）	Architecture Design: Skidmore, Owings & Merrill (SOM)
结构设计：Skidmore, Owings & Merrill（SOM）	Structure Design: Skidmore, Owings & Merrill (SOM)
土木工程：Skidmore, Owings & Merrill（SOM）	Civil Engineering Design: Skidmore, Owings & Merrill (SOM)
高层建筑设计：Skidmore, Owings & Merrill（SOM）	High-rise Architecture Design: Skidmore, Owings & Merrill (SOM)
景观设计：SWA	Landscape Design: SWA
室内设计：Skidmore, Owings & Merrill（SOM）	Interior Design: Skidmore, Owings & Merrill (SOM)
照明设计：Francis Krahe & Associates Inc.	Lighting Design: Francis Krahe & Associates Inc.
幕墙设计：Skidmore, Owings & Merrill（SOM）	Facade: Skidmore, Owings & Merrill (SOM)
合作设计：北京市建筑设计研究院	Joint Design: Beijing Institute of Architectural Design (Architect of Record)
项目地点：中国 · 北京	Location: Beijing, China
项目规模：23 612 m²	Project Scale: 23,612m²
景观面积：116 000 m²	Landscape Area: 116,000m²
建成时间：2017 年	Date of Completion: 2017
设计周期：12 个月	Cycle of Design: 12 months

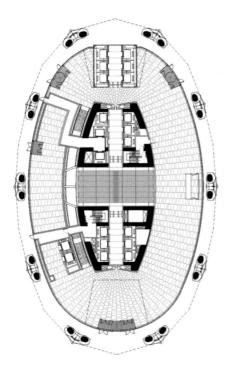
主要大堂层平面
MAIN LOBBY FLOOR PLAN

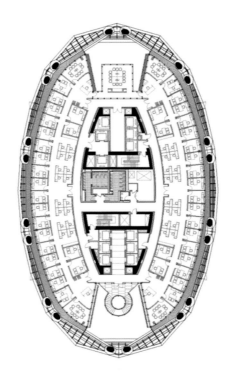
典型中层办公层平面
TYPICAL MIDDLE-LEVEL OFFICE FLOOR PLAN

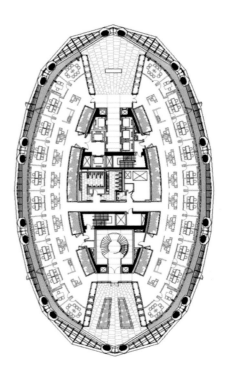
典型高层办公层平面
TYPICAL HIGH-LEVEL OFFICE FLOOR PLAN

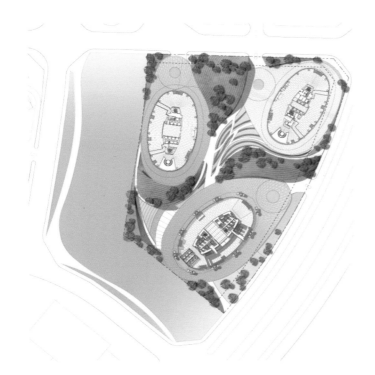

保利国际广场位于北京故宫和首都机场之间,在临近首都机场高速的新建商务区里占据着优越的位置。此处原为北京中央商务区扩建区内的村庄。

项目场地内以一栋主塔楼和两座副塔建筑象征河流内的三块卵石,沿北京的中轴线布置,作为绿化的中心焦点。

三栋大厦的椭圆形楼体基底形状避免了与周围城市结构中硬朗的建筑几何形体的雷同。场地西侧的公园景观被完全融入项目场地,再连通到东侧的公园,形成绵延不断、流畅开放的公共景观。

Located midway between the Forbidden City and Beijing Capital Airport, Poly International Plaza occupies a prominent position in a new business district adjacent to the Capital Airport Expressway. This site used to be a village of the extension area of the Beijing Central Business District.

The project site comprises a main tower and two auxiliary towers which symbolize the three pebbles in the river and are laid out along the central axis of Beijing, becoming a green centre.

The elliptical footprint of the three towers frees the buildings from the rigid geometry of the adjoining urban fabric, allowing the landscape of surrounding parks to continue seamlessly through the project site.

受中国纸灯笼启发，主楼独特的外墙以连续斜肋框架构成宝石切割面般的造型，让玻璃幕墙在反映周围自然和城市景观的同时熠熠生辉。

为了保持椭圆形体的轴向荷载稳定，斜肋框架的构件皆用笔直而非弯曲的部件建造。楼板骨架荷载只传送到斜肋框架的主要节点。因此，每隔一层的楼板骨架都直接接入外框架，而其他层的楼板则用挂件悬挂于上一层的楼板节点上。

Inspired by Chinese paper lanterns, the unique exterior wall of the main building is shaped like a jewel-cut surface with a continuous diagonal rib frame, making the glass curtain wall shining and reflecting the surrounding nature and urban landscape.

To stabilize the axial load of the elliptical body, the components of the diagonal rib frame are made of straight rather than curved parts. Since the load of the floor frame is only transmitted to the main nodes of the diagonal rib frame, the floor frame of every other story is directly connected to the outer frame, while the floor of other stories is hung on the floor nodes of the upper floor with pendants.

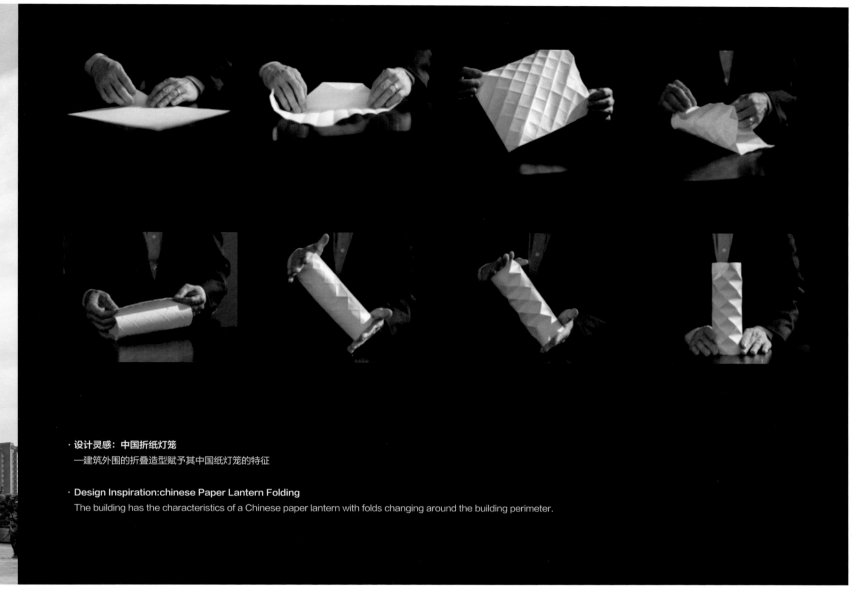

· 设计灵感：中国折纸灯笼
　—建筑外围的折叠造型赋予其中国纸灯笼的特征

· **Design Inspiration:chinese Paper Lantern Folding**
The building has the characteristics of a Chinese paper lantern with folds changing around the building perimeter.

· 设计创新：主动式双层幕墙

· Design Innovation: Active Double-Layer Curtain Wall

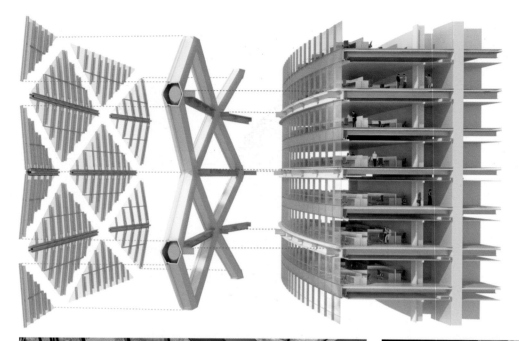

· 新的办公塔楼典范

保利国际广场采用独特的斜肋框架结构体系来构建通高中庭,使之阳光充沛。创新的高度可持续性系统应对了气候和空气质量的挑战。

· A New Paradigm For Office Towers

Poly International Plaza leverages a unique diagrid structural system to introduce full height atria, providing a light-filled spatial experience. The innovative and highly sustainable engineering system addresses climatic and air quality challenges.

丹寨万达小镇
DANZHAI WANDA TOWN

评委点评　JURY COMMENTS

项目处于苗疆腹地，充分挖掘当地建筑特色、民俗文化，多元的非物质文化遗产展示给项目带来了更深层次的底蕴。这种多维度、多层次、多内容的产品设计，使小镇充满活力和趣味，业态设置和建筑形式融合度极高，成为文旅小镇的精品典范。

Located in the hinterland of Miao, this project makes the best of the local architectural features, local culture, and the diverse intangible cultural heritage to present its deep insight. Besides, the design of this project is multi-dimensional, multi-level, and multi-content, making the town dynamic and interesting and the project settings and its architectural forms are perfectly integrated, making it a fine model of cultural tour towns.

第四届地产设计大奖·中国　公建项目 – 金奖
Public Construction Project | Gold Award of the 4th CREDAWARD

业主单位：大连万达集团股份有限公司	Owner: Dalian Wanda Group
建筑设计：万达商业规划研究院有限公司 上海力夫建筑设计有限公司 重庆市设计院	Architectural Design: Wanda Business Planning Research Institute Co., Ltd. Shanghai LIFE Architectural Design Co., Ltd. Chongqing Architectural Design Institute
景观设计：笛东规划设计股份有限公司	Landscape Design: DDON Planning & Design Inc.
幕墙设计：北京和平幕墙工程有限公司	Facade: Beijing Peace Curtain Wall Engineering Co., Ltd.
项目地点：中国·贵州·丹寨	Location: Danzhai, Guizhou, China
项目规模：124 974 m²	Project Scale: 124,974 m²
建筑面积：50 000 m²	Floor Area: 50,000 m²
建成时间：2017 年	Date of Completion: 2017
设计周期：12 个月	Cycle of Design: 12 months

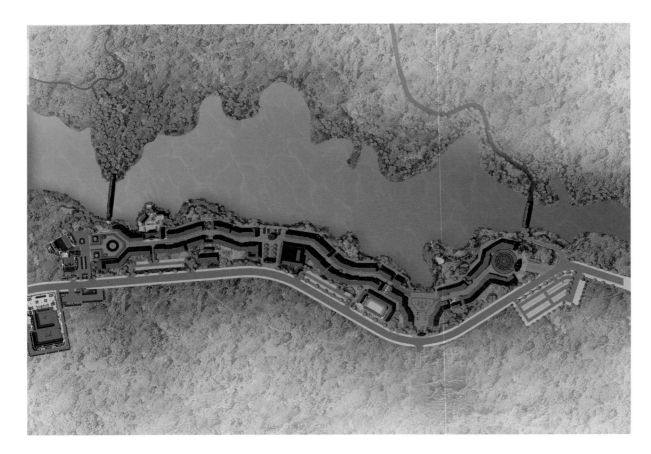

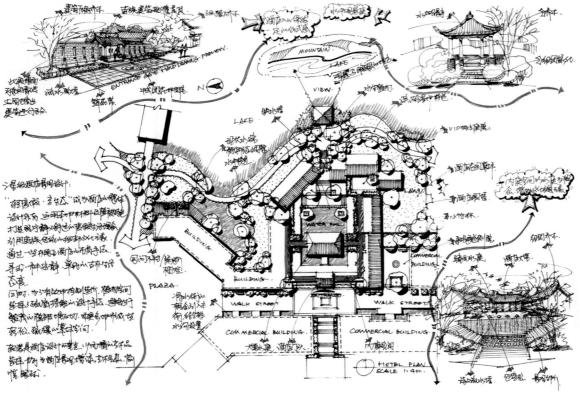

丹寨万达小镇位于贵州黔东南自治州丹寨县东湖西岸，总建筑面积约5万平方米，全长1.5千米。小镇以苗族传统特色建筑为基础，传承丹寨历史文化与民族风情，成功引入7个国家级和17个省级非物质文化遗产项目落户，是一座集"吃、住、行、游、购、娱、教"于一体的文化、非遗、养生旅游小镇，致力于打造成为一个具有全国影响力的苗族非遗文化旅游小镇。

Danzhai Wanda Town is located on the west bank of East Lake, Danzhai County, Qiandongnan Autonomous Prefecture, Guizhou, with a total construction area of about 50,000 square meters and a total length of 1.5 kilometers. Based on the traditional characteristic buildings of the Miao nationality, inheriting the historical culture and ethnic customs of Danzhai, it has successfully introduced 7 national and 17 provincial intangible cultural heritage projects. Combined with culture, intangible cultural heritage and health preservation, Danzhai Wanda Town is a tourist town, providing its visitors with one-stop experience covering foods, stay, traffic, travel, shopping, entertainment and education. Up to now, Danzhai Wanda Town has been a Miao non-legacy cultural tourism town with national influence.

· 总体布局

根据周边环境和地形所作出的理性选择：为了让更多的商铺能够面向东湖，享受湖景，规划采用线性的布局；利用场地的高差形成中间主街和沿湖酒吧街（吊一层）。为了打破街道的沉闷感，设计在中间位置加入了四个广场和三个非遗小院，在北侧相对安静处布置酒店。

· The Overall Layout

The reasonable choice based on the surroundings and topography: A linear layout is used for the planning to make as many stores as possible toward the East Lake to enjoy its view. On one hand, the project makes best use of the height difference of the site to form the main street and the bar street along the lake (add one more floor), on the other hand, places four squares and three intangible-cultural-heritage yards in the street to bring the dull street to life. In addition, to meet its requirements, the hotel lies in the northern quiet place.

在低成本条件下,打造传统苗寨乡土建筑风情是本项目的技术难点,立面中大面积使用的木材是关键。小镇的木纹效果大部分是用木纹涂料来实现的。设计通过对木纹涂料色彩的搭配以及施工工艺的控制,避免了重复、单调的木纹肌理,实现有机自然的整体效果。

It's technically difficult for this project to build the traditional Miao village at a low cost. As to the technical challenges, the key is to massively use wood in the facade. Because the wood grain paint is mainly used to display the wood grain effect in the town, the design focuses on the color matching of wood grain paint and the construction process to avoid repetitive and monotonous wood grain texture and make it natural.

· 2014

约17万人的丹寨全县贫困人口多达4.61万人。

12月1日，万达集团与国务院扶贫办、贵州省扶贫办、丹寨县签订对口帮扶丹寨县"企业包县、整县脱贫"协议，助力丹寨县实现脱贫致富。

Danzhai County with about 170,000 people has as many as 46,100 poor people.

In December, Wanda Group signed an agreement with the Poverty Relief Office of The State Council, the Poverty Relief Office of Guizhou Province and Danzhai County to help Danzhai County lift itself out of poverty.

· 2016

项目5月20日开工建设。万达商业地产技术研发部牵头，与重庆市设计院设计团队联合，将BIM技术首次与特色小镇项目设计结合。

Construction began on May 20. Led by the Technology Research and Development Department of Wanda Commercial Real Estate and the design team of Chongqing Architecture Design Institute, BIM technology was combined with the design of featured town project for the first time.

· 2017

项目7月3日正式运营，启动轮值镇长项目。开业40天，游客量即突破了100万人次大关，历史单日游客最高7.9万人次。

July 3, the official operation, the launch of the rotating mayor project. Within 40 days of its opening, the number of visitors exceeded 1 million, and the highest number of visitors in a single day was 79,000.

·2018

7月3日,开业一周年,全年累计接待游客已达550万人次,是2016年丹寨全县游客数量的600%;丹寨县旅游综合收入达24.3亿元,是2016年全县旅游综合收入的443%,直接带动近2 000人就业,带动全县1.6万贫困人口实现增收。

万达续捐5亿元开展丹寨万达小镇的二期规划建设。小镇剧场及民族文化活动中心开业。

启动扶贫茶园项目,第一期2 000人认领,带动和帮扶丹寨建档立卡贫困户约501人增收。

On July 3, the first anniversary of its opening, the total number of tourists has reached 5.5 million, which is 600% of the total number of tourists in Danzhai County in 2016.The comprehensive tourism income of Danzhai County reached 2.43 billion yuan, 443% of the county's overall tourism income in 2016, directly creating nearly 2,000 jobs and boosting the income of 16,000 poverty-stricken people in the county.

Wanda Group continued to donate 500 million yuan to carry out the second phase of planning and construction, and the town theater and ethnic cultural activity center opened.

The poverty alleviation tea garden project was launched, and the first phase was claimed by 2,000 people, which helped increase the income of about 501 registered poor households in Danzhai.

·2019

扶贫茶园二期上线,启动仪式当天,230亩(约153 410 m²)扶贫茶园被多家企业认领。

4月,贵州省政府宣布丹寨县减贫摘帽,比计划提前两年实现脱贫目标。

6月30日,万达集团宣布,包括研学基地、玻璃栈道、水上游乐、游客集散中心在内的小镇三期4个项目开业。万达捐资3亿元建设的四期项目温泉酒店正式开工,同时启动小镇五期项目规划。

10月17日,2019年全国脱贫攻坚奖表彰大会在北京举行,丹寨小镇项目获得全国脱贫攻坚奖组织创新奖。

The second phase of the poverty alleviation tea garden was launched. On the day of the launch ceremony, 230 mu (153,410 m²) of poverty alleviation tea garden was claimed by many enterprises.

In April, the Guizhou Provincial Government announced that Danzhai County had lifted its poverty reduction target two years ahead of schedule.

On June 30, Wanda Group announced the opening of four projects of the third phase of the town, including a research base, glass walkway, water recreation and a tourist center. Wanda Group donated 300 million yuan for the construction of hot spring hotel, the fourth phase of the project and launched the town's fifth phase project planning.

The Danzhai Town Project won the organizational innovation award of the 2019 National Poverty Alleviation Award in Beijing on Oct. 17.

上海徐汇绿地缤纷城

SHANGHAI XUHUI GREENLAND BEING FUN CENTER

> 评委点评　JURY COMMENTS
>
> 作为 TOD 商业综合体，设计巧妙解决了地铁退让和复杂的人流组织，并创造性打造了多层次屋顶花园，成为城市中的绿洲！
>
> As a TOD commercial complex project, the design delicately lets the subway give a room to it and solves the complicated people flow and creatively builds a multi-level roof garden, making it an oasis in the city!

第四届地产设计大奖 · 中国　公建项目 – 金奖
Public Construction Project | Gold Award of the 4th *CREDAWARD*

业主单位：绿地控股集团 & 上海绿地恒滨置业有限公司	Owner: GREENLAND HOLDING GROUP & Shanghai Greenland Hengbin Properties Limited.
规划设计：株式会社日建设计	Planning Design: NIKKEN SEKKEI LTD
建筑设计：株式会社日建设计	Architectural Design: NIKKEN SEKKEI LTD
联合设计，华东建筑设计研究总院	Joint Design, ECADI
景观设计：mindscape（初步设计）	Landscape Design: mindscape (Development Design);
上海纳千景观环境设计有限公司（施工图设计）	Shanghai Naqian Landscape Design. LLC. (Construction Design)
室内设计：方案设计，Studio Taku Shimizu	Interior Design: Studio Taku Shimizu
中建东方装饰有限公司	China Construction Dongfang Decoration Co., Ltd.
照明设计：bpi	Lighting Design: bpi
幕墙设计：上海凯腾幕墙设计咨询有限公司	Facade: Kighton Facade Consultants Co., Ltd.
摄　　影：mintwow；上海渡影文化传播有限公司	Photographer: mintwow; Shanghai Pdoing Vision & Culture Communication Co., Ltd.
项目地点：中国 · 上海	Location: Shanghai, China
项目规模：304 910 m²	Project Scale: 304,910 m²
建筑面积：21 925 m²	Floor Area: 21,925 m²
建成时间：2017 年	Date of Completion: 2017
设计周期：18 个月	Cycle of Design: 18 months

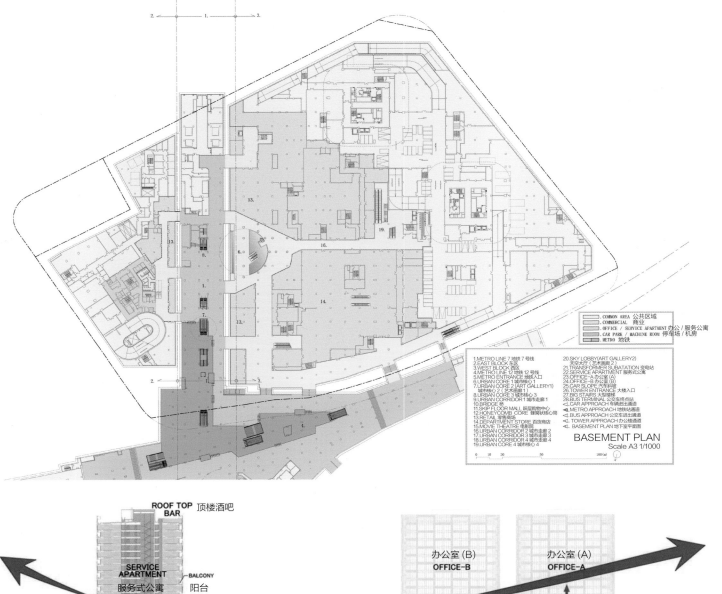

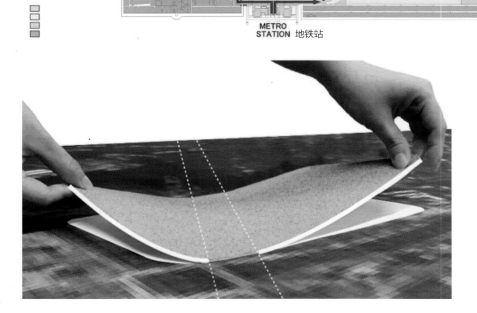

· 地铁站延伸至城市上空的"绿色大地"

上海徐汇绿地缤纷城拥有目前上海最大的开放型屋顶花园，倾斜的大屋顶采用大跨度的结构体系，对于项目实施是前所未有的挑战。

· The Extension Of The Subway Station To The "Greenland" Above The City

With the biggest open roof garden in Shanghai, the massive sloping roof of the building of SHANGHAI XUHUI GREENLAND BEING FUN CENTER uses a large-span structural system, which is an unprecedented challenge for this project implementation.

该项目实现了社会效益和经济效益的双赢,办公物业反馈良好,商场仅开业半年就实现了大幅赢利。

With the good feedback from the office-property management-agency, the project achieved a win-win situation socially and economically, having the substantial profits after opening in only half a year.

上海外滩金融中心
SHANGHAI BUND FINANCE CENTER

> **评委点评　JURY COMMENTS**
>
> 上海外滩金融中心（BFC），来自外滩也必将融入外滩。匠心的打磨使建筑本身与其所属环境形成极致的统一，新旧交融使历史与未来相结合，展现出外滩标志性的全新商业地标。
>
> The Bund Finance Center (BFC) is located at the Bund in Shanghai and melts into the Bund as well. Thanks to the delicate design, BFC and its surroundings harmoniously coexist, marking the new commercial landmark on the Bund.

第四届地产设计大奖·中国　公建项目 – 金奖
Public Construction Project | Gold Award of the 4th CREDAWARD

业主单位：上海证大外滩国际金融服务中心置业有限公司	Owner: Shanghai Zendai Bund Int'l Finance Center Real Estate Co., Ltd.
规划设计：Foster + Partners，赫斯维克工作室	Planning Design: Foster + Partners, Heatherwick Studio
建筑设计：Foster + Partners，赫斯维克工作室	Architectural Design: Foster + Partners, Heatherwick Studio
景观设计：Martha Schwartz Partners	Landscape Design: Martha Schwartz Partners
室内设计：Benoy 贝诺（商业部分）；Kokaistudios（商业部分）	Interior Design: Benoy (Commercial Part) ; Kokaistudios (Commercial Part)
照明设计：bpi	Lighting Design: bpi
项目地点：中国·上海	Location: Shanghai, China
项目规模：426 073 m²	Project Scale: 426,073 m²
建筑面积：426 073 m²	Floor Area: 426,073 m²
建成时间：2017 年	Date of Completion: 2017
设计周期：48 个月	Cycle of Design: 48 months

外滩金融中心是一个临近上海滨水区的新的、重要的综合用途开发项目，由 Foster + Partners 和 Heatherwick Studio（赫斯维克工作室）共同设计。该建筑占据了上海外滩一块引人注目的场地，对上海最著名的街道的"终点"作了新的诠释。总体规划考虑了行人的高度流动性，将其构思为老城和新金融区之间的一个连接点。两座180米高的地标性塔楼的设计灵感源于都市环境，塔楼位于场地南部，滨水大楼高低不同、错落有致、规模相当，展示了19世纪外滩宏伟的地标性建筑的节奏感。

42万平方米的8座大楼的开发将高档办公楼与精品酒店、文化中心和各类豪华零售空间融为一体，一切均围绕景观公共广场进行布局。办公楼以独特的"企业大厦"为支撑——这是一个商业合作伙伴的精英网络平台。零售空间为垂直分层布置，涵盖了精品店、国际品牌概念店、豪华购物中心和餐馆。精心制作的石材和青铜件为大楼增添了宝石般的质感。每个体块边缘均由富于质感的手工制作的花岗岩制成，随着大楼的升高，逐渐变得纤细，给人以基座坚实、顶部通透的感觉。

本方案的核心是一个灵活的艺术文化中心，将展览和活动大厅与表演场所融为一体，灵感源于中国传统剧院的露天舞台。该中心致力于成为一个国际艺术与文化交流平台、品牌活动、产品发布和企业运营场所。大楼周围环绕着一个可移动遮蔽物，可变换用途，展现阳台上的舞台以及朝向浦东的美景。

The Bund Finance Center (BFC) is a major new mixed-use development project which is close to the waterfront, jointly designed by Foster+Partners and Heatherwick Studio. The BFC occupies a prominent site on the Bund and redefines the "end point" of the most famous street in Shanghai. Considering the dense pedestrians, the project is designed as a connecting point between the old town and the new financial district. Inspired by the urban context, the two 180-metre landmark towers with different heights and suitable scale are placed in the south of the site, presenting the rhythm of the grand landmark buildings on the Bund in the nineteenth-century.

With the total area of 420,000 m^2, the eight towers roll the premium offices, the boutique hotel, the cultural center and the various luxury shops into one. Besides, all of them are arranged around the scenic public plaza. The main function of the office building serves as a "Corporate Mansion" -a networking platform for business elites. The retail spaces are vertically layered with boutiques, concept stores for international brands, a luxury shopping mall and restaurants. The delicate stones and bronze add a jewel-like quality to the buildings. The edges of each building, which are made of richly textured, hand-crafted granite, become slimmer as they rise, giving the impression of solid base and transparent top.

Inspired by the open stages of traditional Chinese theatres, the heart of the project is a flexible arts & culture center, which combines exhibition and events with show. The center is designed as a place for international arts & culture exchange, brand marketing, product launching and business operation. The building is encircled by a moving veil, which adapts to the changing use of the building and reveals the stage on the balcony and the views towards Pudong.

平安金融中心
PING AN FINANCE CENTER

" 评委点评 JURY COMMENTS

作为中国最高建筑之一，平安金融中心成功地解决了诸多挑战性的技术难题，为深圳创造了一个新的城市地标。

The tower of Ping An Finance Center is one of the highest buildings in China, the designers have successfully solved many challenging technical problems and created a new urban landmark for Shenzhen.
"

第五届地产设计大奖·中国 公建项目 – 金奖
Public Construction Project | Gold Award of the 5th CREDAWARD

业主单位：平安保险公司	Owner: Ping An Insurance Company
建筑设计：KPF 建筑设计事务所	Architectural Design: Kohn Pedersen Fox Associates(KPF)
本地设计单位，悉地国际	Local Design Institute, CCDI
景观设计：AECOM	Landscape Design: AECOM
室内设计：KPF 建筑设计事务所（办公）	Interior Design: Kohn Pedersen Fox Associates(KPF) (Office)
J&A（塔楼）	J&A (Tower)
Benoy 贝诺（商业）	BENOY (Retail)
照明设计：LPA	Lighting Design: Lighting Planners Associates
幕墙设计：ALT	Facade: ALT
项目地点：中国·深圳	Location: Shenzhen, China
项目规模：462 000 m²	Project Scale: 462,000m²
建筑面积：388 680 m²	Floor Area: 388,680m²
建成时间：2016 年	Date of Completion: 2016
设计周期：48 个月	Cycle of Design: 48 months

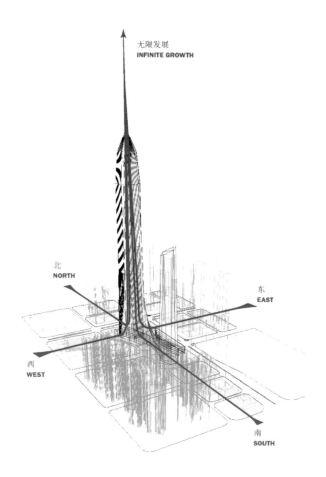

平安金融中心位于深圳日益发展的中央商务区，是该区域标志性的同时也是实际的中心。项目拥有100多层办公空间以及设有零售商业和会议空间的大型裙楼，并与周边商业及住宅地产和公共交通设施相连。

塔楼从基地拔地而起，采用不锈钢和玻璃立面，确定了项目的主要基调。大楼的四个立面被三角形的不锈钢垂直肋包围，从建筑的石材基底向上延伸。顶层设有空中大堂，其如今成为了一个空中餐厅，在这个空间内可以感受到阳光从四面照射进来。

Ping An Finance Center is the physical and iconic center of Shenzhen's growing central business district. With more than 100 floors of office space and a large podium with retail and conference space, the project also connects to neighboring commercial and residential properties and public transportation.

The stainless steel and glass cladded tower rises from the site, anchoring the development. Its four facades are sheathed in triangular-shaped stainless steel vertical piers that extend from the building's stone base. At the top is a sky hall which serves as a restaurant in the sky, flooded with daylight from all sides.

· 稳定支撑，庞大体量

塔楼纵向由 8 根巨柱支撑，深灰色花岗岩地基从视觉上展现庞大的体量，传达巨大的力量感和稳定性。

· **Stable Support, Huge Physical Volume**

The tower is supported by eight huge columns in the longitudinal direction. In addition, the dark gray granite base visually shows the substantial scale, conveying a sense of great strength and stability.

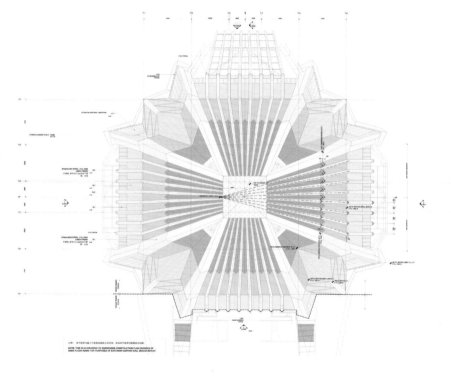

· 雷电防护"网"设置

可回收的外立面材料极具环保性,外立面 V 形钢制支柱形成一张防护"网",收集侧面雷电,全面防护平安金融中心。

· Thunder And Lightning Protection

Not only are the recyclable materials of the facade extremely environment-friendly, the V-shaped steel piers on the facade take a role as a protective "net" to collect the side lightning to fully protect the tower of Ping An Finance Center.

· 强力建材，减少负荷
  支撑结构有效缓解空中受力点的张力，裙楼采用玻璃和石材覆层，边柱和八层对角钢支撑整体结构。

· 标志性中心
  塔楼锥形外立面减少了 40% 的风负荷，让平安金融中心成为新中央商务区的标志性中心。

· Strengthening Building Materials, Reducing Loads
  The supporting structure effectively relieves the tension of the stress points in the air. In addition, the podium is covered with glass and stone, and the side pillars and eight-story angle steel support the overall structure.

· Landmark
  The tapered facade of the tower reduces the wind load by 40%, making the tower of Ping An Finance Center become the landmark of the new central business district.

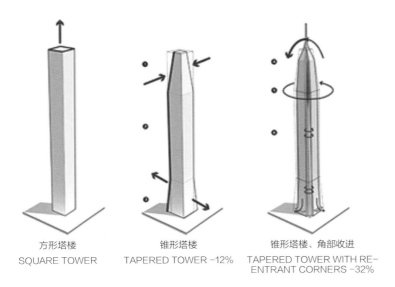

| 方形塔楼 | 锥形塔楼 | 锥形塔楼、角部收进 |
| SQUARE TOWER | TAPERED TOWER -12% | TAPERED TOWER WITH RE-ENTRANT CORNERS -32% |

与方形塔楼相比，锥形形体的倾覆力矩比规范要求减少约 32%。
The tapered shape has a 32% reduction in overturning moment compared to a typical square shape.

倾覆力矩
OVERTURNING MOMENT

倾覆力矩比较
OVERTURNING MOMENT COMPARISON

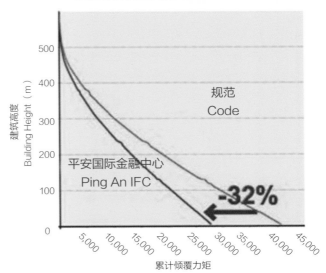

| 方形塔楼 | 锥形塔楼 | 锥形塔楼、角部收进 |
| SQUARE TOWER | TAPERED TOWER -12% | TAPERED TOWER WITH RE-ENTRANT CORNERS -35% |

与方形塔楼相比，锥形塔楼的风荷载比规范要求减少约 35%。
The corner-retracted shape has a 35% reduction in the wind load compared to a typical square shape.

风荷载
WIND LOAD

风荷载比较
WIND LOAD COMPARISON

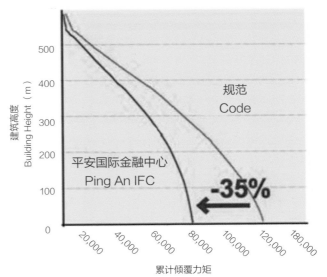

周大福金融中心
CTF FINANCE CENTER

> 评委点评　JURY COMMENTS
>
> 作为超高层综合体，创新性地解决了办公、公寓、酒店不同功能的组合，其幕墙细部设计独具一格。
>
> As a super high-rise complex building, it innovatively realizes the combination of different programs including office, apartment and hotel. In addition, the detail design of its curtain wall is unique.

第五届地产设计大奖·中国　公建项目 – 金奖
Public Construction Project | Gold Award of the 5th CREDAWARD

业主单位：新世界中国	Owner: New World China
建筑设计：KPF 建筑设计事务所	Architectural Design: Kohn Pedersen Fox Associates(KPF)
本地设计单位，广州市设计院	Local Design Institute, Guangzhou Design Institute
执行建筑机构，利安顾问有限公司	Executive Institute, Leigh & Orange Limited.
景观设计：PLA	Landscape Design: PLA
室内设计：Yabu Pushelberg（酒店）	Interior Design: Yabu Pushelberg (Hotel)
Perception Design（服务式公寓部分）	Perception Design (Serviced Apartments)
CallisonRTKL（商业部分）	CallisonRTKL (Retail)
照明设计：LPA（室外）	Lighting Design: LPA (Exterior)
Isometrix Lighting Design（室内）	Isometrix Lighting Design (Interior)
项目地点：中国·广州	Location: Guangzhou, China
项目规模：508 000 m²	Project Scale: 508,000 m²
建筑面积：390 800 m²	Floor Area: 390,800 m²
建成时间：2016 年	Date of Completion: 2016
设计周期：72 个月	Cycle of Design: 72 months

周大福金融中心作为中国最高的建筑之一，是超塔设计展现对当地文化、环境以及城市肌理敏感度的杰出案例。

周大福金融中心塔楼采用层层退台的形式，与该综合体塔楼的办公、服务式公寓和酒店功能相呼应。退台形成的空中露台提供了极致高度下的户外体验。带有冰裂纹图案的釉面陶土板在阳光的照射下使外墙呈现亮白色闪耀的色泽。同时，竖向布置的陶板与立面照明和可开启通风口融为一体。

As one of the China's tallest structures, the CTF Finance Center demonstrates the alignment of supertall design with cultural, environmental, and contextual sensitivity.

The CTF Finance Center tower form consists of a stepped volume corresponding to the different programs within the mixed-use tower: Office, serviced hotel and hotel. These stepped volumes serve as sky terraces providing outdoor experiences at extreme height. Glazed terracotta panels with crackle pattern are used to give a white vertical shimmering effect to the exterior wall under the sun. The vertical panel is also integrated with facade lighting and operable vent.

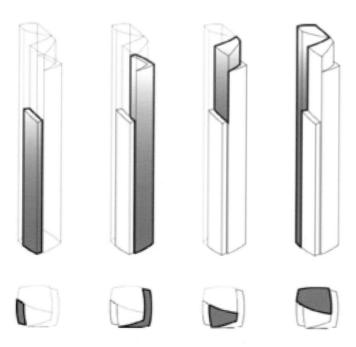

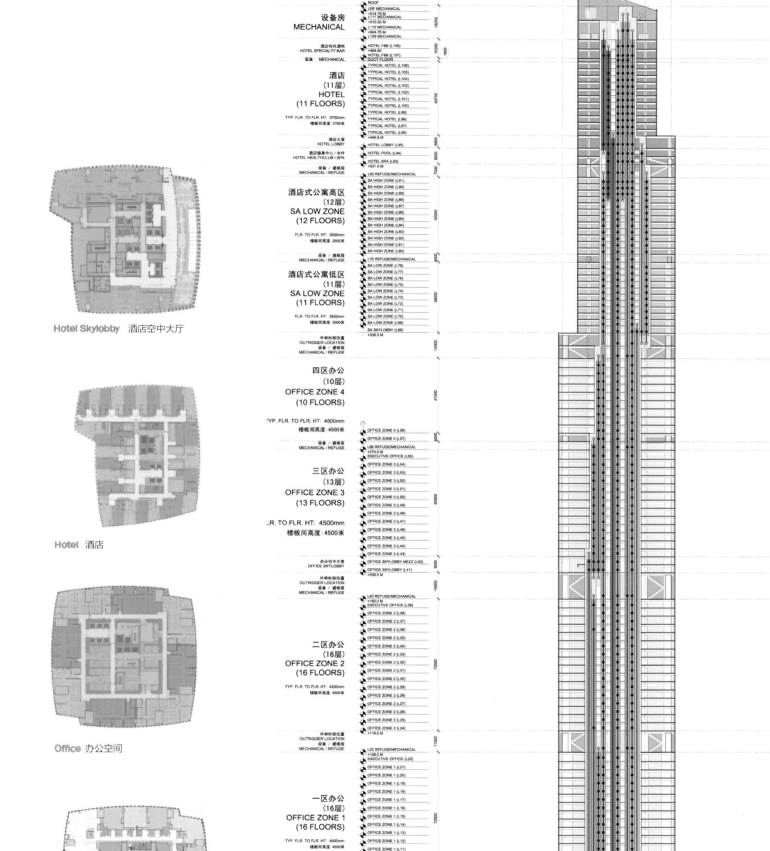
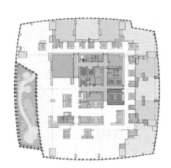

周大福金融中心定位为向上攀升至高空的水晶造型结构,设有陶板外立面,体现该材质历史特征。亮白的饰面与旁边的广州国际金融中心(西塔)的深沉颜色形成对比。幕墙上采用上釉陶土板作为窗间壁,既保证了每层的通透视野,同时提供充足的遮阳功能,保证在广州这样的亚热带气候下的自然通风。建筑采用了可持续性措施,例如高效冷却器和热回收。

Positioned as a crystalline form that ascends to the sky, the CTF Finance Center features a terracotta facade attuned to the material's history. Its bright finish creates a contrast to the dark color of adjacent Guangzhou IFC tower. The facade's ceramic-clad piers preserve floor-to-ceiling views and offer generous shading to the exterior, providing natural ventilation in Guangzhou's tropical climate. The building employs additional sustainability measures, like high-efficiency chillers and heat recovery.

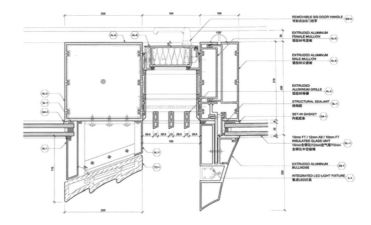

③ TYPICAL VERTICAL MULLION DETAIL @VISION @WALL TYPE A
标准竖向竖框节点 @ 可视区 @ 墙型 A

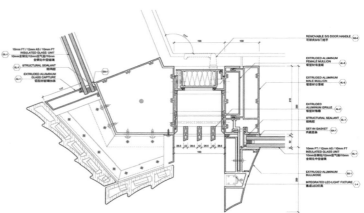

① TYPICAL VERTICAL OUTSIDE CORNER DETAIL @VISION @WALL TYPE A
标准竖向外转角节点 @ 可视区 @ 墙型 A

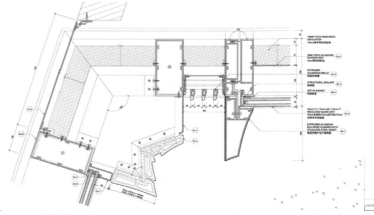

① TYPICAL VERTICAL INSIDE CORNER DETAIL @VISION @WALL TYPE A&B
标准竖框内转角节点 @ 可视区 @ 墙型 A&B

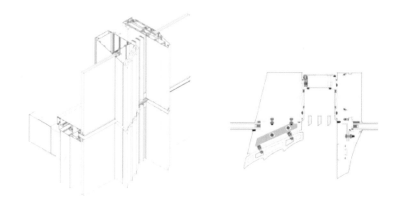

船厂 1862
MIFA 1862

评委点评 JURY COMMENTS

本设计的出彩之处就在于：把似乎是陈旧的片段表皮与隐匿在后代表时代的新功能平静地组合起来，表现出了对于城市文脉的思考。

The highlight of the design is that it skillfully combines the seemingly old, fragmented appearance with the new functions which are hidden in the post-age, showing the way of thinking of the urban context.

第五届地产设计大奖·中国 公建项目 – 金奖
Public Construction Project | Gold Award of the 5th CREDAWARD

业主单位：中船置业有限公司	Owner: Shanghai Chime Shipping Property Ltd.
规划设计：上海市城市规划设计研究院 　　　　　Gensler	Planning Design: Shanghai Urban Planning Design Research Institute 　　　　　　　　　Gensler
建筑设计：隈研吾建筑都市设计事务所	Architectural Design: Kengo Kuma and Associates
景观设计：隈研吾建筑都市设计事务所	Landscape Design: Kengo Kuma and Associates
室内设计：隈研吾建筑都市设计事务所	Interior Design: Kengo Kuma and Associates
照明设计：奥雅纳（上海）	Lighting Design: ARUP Shanghai
联合设计：上海建筑设计研究院有限公司	Joint Design: Shanghai Architectural Design & Research Co., Ltd.
幕墙设计：奥雅纳（上海）	Facade: ARUP Shanghai
项目地点：中国·上海	Location: Shanghai, China
项目规模：31 626 m²	Project Scale: 31,626 m²
建筑面积：26 010 m²	Floor Area: 26,010 m²
建成时间：2017 年	Date of Completion: 2017
设计周期：29 个月	Cycle of Design: 29 months

摄影师：Eiichi Kano
Photographer: Eiichi Kano

约198米 (about 198 m)

老厂房原平面图
The original floor plan of the old factory

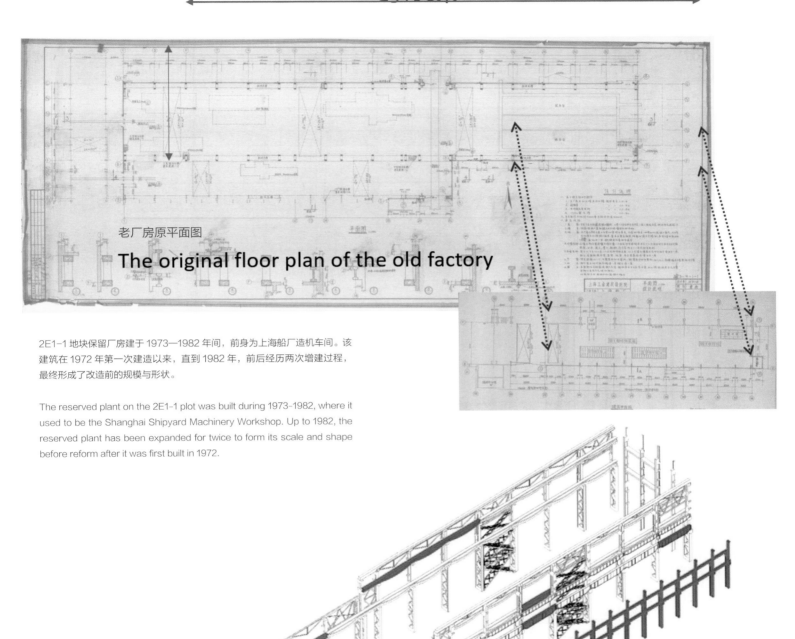

2E1-1 地块保留厂房建于 1973—1982 年间，前身为上海船厂造机车间。该建筑在 1972 年第一次建造以来，直到 1982 年，前后经历两次增建过程，最终形成了改造前的规模与形状。

The reserved plant on the 2E1-1 plot was built during 1973-1982, where it used to be the Shanghai Shipyard Machinery Workshop. Up to 1982, the reserved plant has been expanded for twice to form its scale and shape before reform after it was first built in 1972.

红色为拆除部分
The dismantled part marked in red

项目前身为 198 米长 45 米宽的工业厂房，分为高低两跨。高跨檐口高度为 24 米，宽度为 30 米；低跨檐口高度为 14 米，宽度为 15 米。结构为混凝土格构柱，鱼腹式预应力吊车梁，预应力混凝土屋架，预应力混凝土屋面板。

With a length of 198 meters and a width of 45 meters, the industrial plant is divided into high span and low one, the cornice of the high span is 24 meters long and 30 meters wide, the cornice of the low-span is 14 meters long and 15 meters wide. The structure of the plant is concrete lattice columns, fish-belly prestressed crane beams, prestressed concrete roof trusses, and prestressed concrete roof slabs.

摄影师：Eiichi Kano
Photographer: Eiichi Kano

· **色彩混杂的砖块造就建筑质感的再生**

老船厂色彩混杂的砖块造就了强烈的质感层次，4种不同颜色的砖块随机打乱排布，通过不锈钢固定件连接，形成一面具有通透性的砖幕墙。这面随机留出孔隙，像粒子一样的砖幕墙与原有砖墙的粗糙、色彩繁杂达到了协调统一。

· **Mixed-Colored Bricks Create The Regeneration Of Architectural Texture**

The mixed-colored bricks of the Shanghai Shipyard create a strong texture level. The bricks with four different colors are selected and laid out by chance to form a transparent brick curtain wall with random pores. The brick curtain wall is like particles, making it harmoniously match the roughness and complex colors of the original brick wall.

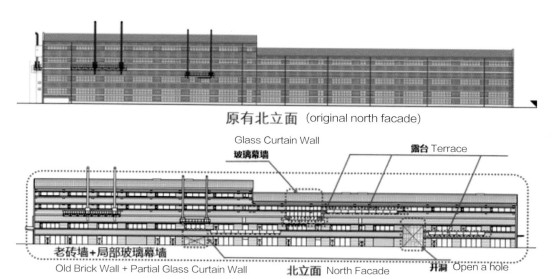

原有北立面 (original north facade)

老砖墙+局部玻璃幕墙 Old Brick Wall + Partial Glass Curtain Wall　北立面 North Facade　井洞 Open a hole
玻璃幕墙 Glass Curtain Wall　露台 Terrace

朝向黄浦江（过去）：保留沿江城市工业记忆，最大限度保留原有立面

Towards the Huangpu River (Past): Preserving the industrial memory of the city along the river, keeping the original facade to the utmost extent.

原有南立面 (original sorth facade)

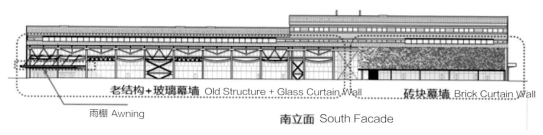

老结构+玻璃幕墙 Old Structure + Glass Curtain Wall　砖块幕墙 Brick Curtain Wall
雨棚 Awning　南立面 South Facade

朝向黄浦江（未来）：暴露立式结构，玻璃幕墙后衬，局部悬挂砖幕墙

Towards Huangpu River (Future): Exposing the vertical structure, placing the lining behind the reflection glass curtainwall, partly hunging the brick curtainwall

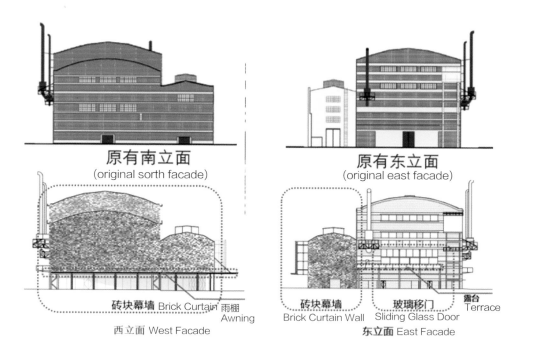

原有南立面 (original sorth facade)　原有东立面 (original east facade)

砖块幕墙 Brick Curtain　雨棚 Awning　砖块幕墙 Brick Curtain Wall　玻璃移门 Sliding Glass Door　露台 Terrace
西立面 West Facade　东立面 East Facade

摄影师：Erieta Attali

Photographer: Erieta Attali

- **特色悬挂陶砖墙**
  悬索吊挂结构
  4 种颜色砖块进行由密到疏的随机组合
  与历史文脉相呼应
  室内形成特色的光影效果
  起到部分隔热作用

- **The Hanging Ceramic Brick Walls With Characteristics**
  Suspension structure
  In terms of the decreasing density, four-color bricks are randomly laid out.
  Echoing with historical context
  Forming a distinctive light and shadow effect inside the room
  Partly insulating heat

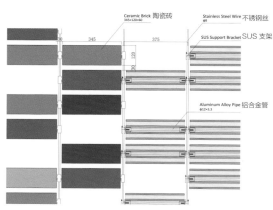

砖屏立面详图　BRICK SCREEN ELEVATION DETAIL

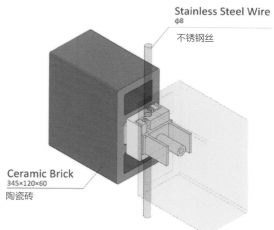

注：图中数字单位均为 mm。

摄影师：Eiichi Kano
Photographer: Eiichi Kano

摄影师：Eiichi Kano
Photographer: Eiichi Kano

腾讯总部
TENCENT HEADQUARTERS

> **评委点评　JURY COMMENTS**
>
> 创新的总部办公设计理念在垂直空间中提供了交流与沟通的场所，为未来办公场所设计提供了又一种可能。
>
> The innovation design of the headquarters provides a place for communication in the vertical space, and also provides another possibility for the future office design.

第五届地产设计大奖·中国　公建项目－金奖
Public Construction Project | Gold Award of the 5th CREDAWARD

业主单位：腾讯控股有限公司	Owner: Tencent Holdings Limited.
建筑设计：NBBJ 建筑设计事务所	Architectural Design: NBBJ
景观设计：NBBJ 建筑设计事务所	Landscape Design: NBBJ
幕墙设计：宋腾添玛沙帝；英海特集团	Facade: Thornton Tomasetti; Inhabit Group
摄　　影：Tim Griffith, 邵峰	Photographer: Tim Griffith, Shao Feng
结构顾问：AECOM	Structural Consultant: AECOM
机电顾问：纽约 WSP 集团	Mechanical / Electrical Consultant: WSP Group, New York
国内设计院：深圳市同济人建筑设计有限公司	Associate Architect: Shenzhen Tongji Architects Co., Ltd.
交通顾问：奥雅纳	Traffic Consultant: Arup
垂直交通顾问：奥雅纳	Vertical Transportation Consultant: Arup
项目地点：中国·深圳	Location: Shenzhen, China
项目规模：18 700 m²	Project Scale: 18,700 m²
建筑面积：345 570 m²	Floor Area: 345,570 m²
景观面积：350 000 m²	Landscape Area: 350,000 m²
建成时间：2017 年	Date of Completion: 2017
设计周期：14 个月	Cycle of Design: 14 months

腾讯总部由三座"空中街道"将两栋塔楼衔接起来,并经由这些空中连接层引入多功能空间,包括社区活动空间、花园和康体中心等。这样的环境使员工变得更健康、更加富有创意活力。此外,总部的空中连桥还提供了屋顶公园,为来自不同部门的员工带来中央聚集场所。

Tencent Headquarters features the two towers connected by the three sky streets, which offer multi-function spaces such as community areas, gardens and fitness center. It is no wonder that the employees become healthier and more creative in this environment. In addition, the headquarters' skybridge provides the rooftop park that offers the central gathering space for employees from different departments.

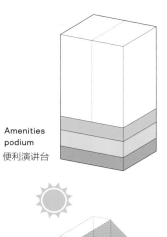

20th-century office tower typology
二十世纪的办公大楼类型

Amenities podium
便利演讲台

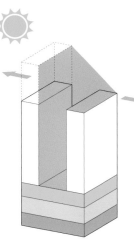

Floor plates narrowed for light and views
让地板变窄以提供光线和视野
South tower raised to shade north tower
升起南塔以遮蔽北塔

Knowledge link:
Education area, conferencing area, library
知识链接：教学区、会议室、图书馆
Health link:
Fitness facilities
健康链接：健身设施
Culture link:
Exhibition hall, auditorium, cafés shop
文化链接：展览厅、礼堂、咖啡厅

Amenities distributed for a networked culture
网络文化的便利设施分布

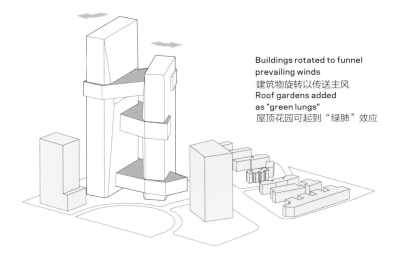

Buildings rotated to funnel prevailing winds
建筑物旋转以传送主风
Roof gardens added as "green lungs"
屋顶花园可起到"绿肺"效应

· 追求革新变化
腾讯总部彰显了独特的设计创新理念，对新一代的高层建筑进行了重新定义，同时也为城市建设及其他企业树立了发展典范。

· 高效融合的商务办公
腾讯总部将一整栋高层建筑划分为 50 层和 39 层的双塔楼，通过三条分别代表着文化、健康、知识的连接层衔接到一起。

· Pursuing Innovation And Change
The design of Tencent Headquarters illustrates the unique design innovation that redefines the next-generation skyscraper, while serving as a model for city planning and other company planning.

· Efficient And Integrated Business Office
The headquarters links two towers, one is 50-stories and the other is 39-stories. It uses three connective layers to promote culture, health and knowledge.

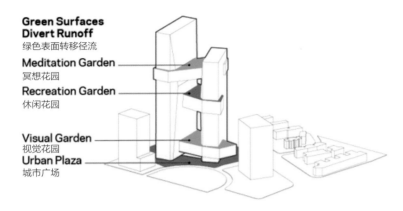

70% of storm water captured
收集 70% 的雨水

40% less energy used compared to other high rise towers
与其他高层塔楼相比，能耗减少 40%

Green Surfaces Divert Runoff
绿色表面转移径流
Meditation Garden
冥想花园
Recreation Garden
休闲花园
Visual Garden
视觉花园
Urban Plaza
城市广场

与传统高层建筑相比，本项目采用被动节能策略将碳排量和能耗量减少了 40%。为达到节能目标，两栋腾讯塔楼的方位设置均有助于减少日照热量。与此同时，设计还利用可开启窗获得室外新风，从而实现能耗节省并提升空气质量。

Compared with traditional high-rise buildings, this project uses passive energy saving strategies to cut down carbon emissions and energy consumption by 40%. To reach that goal, the orientation of the two Tencent towers reduce heat from the sun, meanwhile, operable windows allow fresh air inside to reduce energy use and enhance air quality.

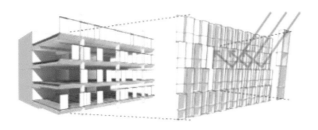

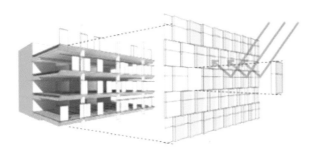

为应对洪水和暴雨等气候变化影响，腾讯总部积极呼应海绵城市计划，成为深圳市首批满足海绵城市发展标准的总部项目。腾讯总部的两座塔楼的设计最大限度地利用了隔热保温与可开启窗。此举每年能够节约近 100 万美元的能耗成本，还有助于打造更健康的环境。

腾讯总部采用的被动外立面系统能随着日照角度的变化而变化，使得每个立面都能顺应不同的日照强度。此外，建筑外墙采用先进的幕墙材料，有助于发挥隔热效应并尽可能减少眩光。

To address impact of climate change, such as, floods and rainstorms, Tencent Headquarters is the first to meet the standards of China's Sponge City Initiative in Shenzhen. Its two towers are designed to make the best of the thermal insulation as well as operable windows. This lowers energy costs by nearly one million dollars per year and creates a healthier workplace as well.

The headquarters' passive facade system can vary with the arc of the sun, allowing each facade to adapt to different light intensities. In addition, the exterior wall uses advanced curtain wall materials to insulate the building from heat and minimize glare.

02 居 住 项 目 RESIDENTIAL PROJECT

01　万科良渚文化村未来城二期桂语里 —————————————————— 158
　　　VANKE LIANGZHU NEW TOWN FUTURE LIFE PHASE II – GUIYULI

02　北京壹号院 ———————————————————————————— 168
　　　BEIJING ONE SINO PARK

03　重庆龙湖舜山府 ——————————————————————————— 174
　　　CHONGQING LONGFOR SHUNSHAN MANSION

182	君山生活美学馆 JUNSHAN CULTURAL CENTER	*04*
190	瀚海晴宇 HANHAI LUXURY CONDOMINIUMS	*05*
198	上海仁恒公园世纪 SHANGHAI YANLORD PARK CENTURY	*06*

万科良渚文化村未来城二期桂语里

VANKE LIANGZHU NEW TOWN FUTURE LIFE PHASE II – GUIYULI

> **评委点评　JURY COMMENTS**
>
> 作为万科原创性微墅产品，项目对于极小墅居和外部空间的研究开国内先河、独具匠心，在居住舒适度、小面积、经济性之间获得良好的平衡。巷弄曲折、粉墙黛瓦、竹影婆娑的江南意境，由园入坊、由坊进巷、由巷通宅的空间序列，老杭州的写意生活就这样徐徐展开。
>
> As Vanke's original project, the micro villas, the project is the first and unique when it comes to researching the extremely small villas and external space. On the one hand, the project keeps a good balance among comfortable experience, a small living area and costs. On the other hand, with the artistic conception of the south of the Yangtze River, which is shown in twisting and turning alleys, white walls, black tiles, and bamboo shadows, the life of old Hangzhou is slowly unfolded in terms of the sequence of spaces from the garden to the square, from the square to the alley, and from the alley to the house.

第三届地产设计大奖·中国　居住项目 – 金奖
Residential Project | Gold Award of the 3rd *CRED*AWARD

业主单位：万科集团	Owner: Vanke Group
规划设计：AAI 国际建筑师事务所	Planning Design: ALLIED ARCHITECTS INTERNATIONAL
建筑设计：AAI 国际建筑师事务所	Architectural Design: ALLIED ARCHITECTS INTERNATIONAL
景观设计：杭州午人景观设计	Landscape Design: 5renz Landscape
室内设计：金螳螂	Interior Design: Gold Mantis
项目地点：中国·杭州	Location: Hanzghou, China
项目规模：23 000 m²	Project Scale: 23,000 m²
建筑面积：800 000 m²	Floor Area: 800,000 m²
景观面积：6 300 m²	Landscape Area: 6,300 m²
建成时间：2016 年	Date of Completion: 2016
设计周期：5 个月	Cycle of Design: 5 months

桂语里用杭州老底子坊巷格局，打造了一个"骨子里的江南"，是杭州万科推出的一款创新微墅产品，容积率接近1.0，套均面积控制在140平方米，实现了"有天有地"的墅式生活。规划布局借鉴了杭州的老式坊巷，采用传统民居宅第类建筑"多进＋多跨＋园宅组合"的构成模式。作为一款微墅产品，它注重面积的控制，同时通过精细化设计，在居住舒适度和小面积间找到良好的平衡。

The mini villas created by Hangzhou Vanke, also known as Guiyuli, display a classical "Jiangnan" style by using the traditional structure of the old lanes and alleys in Hangzhou. With near 1.0 Floor Area Ratio(FAR), the average area of a single house is 140 m², achieving a life style of owning sky view and green courtyards. The project references the traditional structure of the old lanes and alleys in Hangzhou, and adopts the traditional style of the residential buildings to build the house with the model of "Multi-entry+Multi-span+Garden". As a mini villa, the project stresses controlling its area, meanwhile, keeping a good balance between living comfort and small area through the refined design.

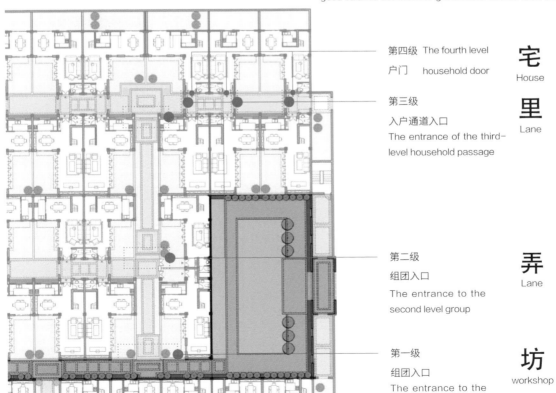

第四级 The fourth level
户门 household door

宅 House

第三级
入户通道入口
The entrance of the third-level household passage

里 Lane

第二级
组团入口
The entrance to the second level group

弄 Lane

第一级
组团入口
The entrance to the first level group

坊 workshop

四级入口，给组团的空间带来节奏变化。

The four-level entrance brings the changing rhythm to the space of the group.

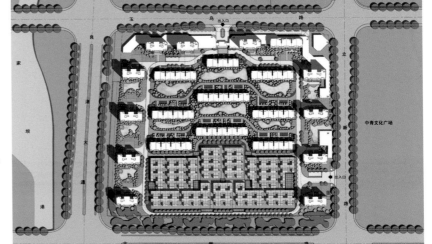

· 小而美的创新别墅

拥有独立庭院的别墅是众多置业者的梦想。但在土地价值高涨的背景下,同时由于低容土地的稀缺,别墅通常有昂贵的价格。小面积、高强度的别墅产品,可有效控制价格,让更多的置业者圆梦。

· Small And Beautiful Innovative Villa

Villas with independent courtyards are many home buyers' dreams. However, villas are usually expensive due to the high price of land and the scarcity of low-capacity land, thus making small-scale and high-strength villas with reasonable prices more popular.

· 创新产品——"乐高"微墅

选用"乐高"作为产品的代号,寓意是简单的基本单元经过平面组合、立体堆叠后,可以实现丰富的变幻。

乐高首先是标准的用地,户型单元均为L形平面,两个户型单元通过二层进行连接,首层中间为入户通道,形成了一个基本模块。基本模块可以东西、南北拼接成更大范围的单元,形成居住的组团。由于拼接的灵活和多联,有效节约了宅间空间,提高了土地使用效率。

· Innovative Product ——"Lego" Villa

The "Lego", the code name of this product, means that the simple basic unit can be rich and colorful via plane combination and 3D stack.

First, the Lego villas choose a standard land. The user's units are all L-shaped. The two units are connected through the second floor. In the middle of the first floor is the entrance passage to form a basic module, which can be spliced into a larger unit from east to west and from north to south to form a residential cluster to effectively improve the land use.

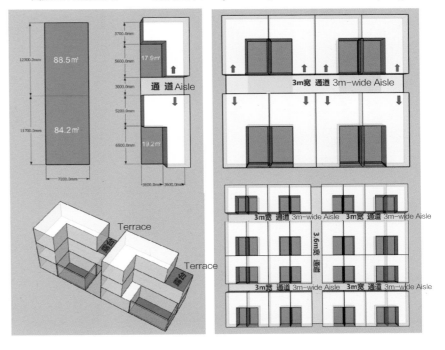

北京壹号院
BEIJING ONE SINO PARK

评委点评　JURY COMMENTS

设计力求突破中国奢华住宅的形式窠臼，选择运用富有现代感的建筑材料和建构技艺塑造住区的整体形象，项目的成功实践对于中国高端居住模式探索和居住空间的创新具有重要意义。

The design strives to break through the stereotypes of the Chinese luxury residences, choosing the modern building materials and construction techniques to shape the overall image of the residential area. The successful practice of the project takes a significant role to explore the high-end residential model and the innovation of residential space in China.

第三届地产设计大奖·中国　居住项目－金奖
Residential Project | Gold Award of the 3rd CREDAWARD

业主单位：融创中国	Owner: SUNAC
规划设计：goa 大象设计	Planning Design: GOA
建筑设计：goa 大象设计	Architectural Design: GOA
景观设计：香港贝尔高林景观设计公司	Landscape Design: Belt Collins International (HK) Limited.
室内设计：李玮珉建筑师事务所	Interior Design: LWM ARCHITECTS
M+W	M+W
摄　　影：舒赫，© 是然建筑摄影	Photographer: SHU HE, ©Schran Image
项目地点：中国·北京	Location: Beijing, China
建筑面积：89 000 m²	Floor Area: 89,000 m²
建成时间：2017 年	Date of Completion: 2017
设计时间：48 个月	Cycle of Design: 48 months

北京壹号院位于北京核心地带，毗邻北京农展馆公园，是都市中心难得的宁静栖居之地。开发的目标在于营造一处北京城内的现代精品社区，让每位居住者都能享受到极致诗意的居住体验。规划布局以充分利用资源优势为原则，场地西侧留出开阔的景观面并依据每栋楼的布局差异作出针对性的设计，使客厅、餐厅等生活空间更多地面向农展馆公园，让优越的景观资源最大化地渗透于居住之中。

Located in the core of Beijing, Beijing One Sino Park, which is adjacent to the park of the National Agricultural Exhibition Hall, is a precious place for people to enjoy the peaceful life in Beijing. The design aims to create a modern boutique community in the city of Beijing, making its residents enjoy the poetic life. The tailored design makes best of the resources, leaving an open landscape on the west side to make the living spaces (such as a living room, a dining room)face the park to enjoy the best views as many as possible.

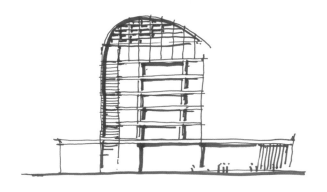

总平面图
Overall plan

· 突破中国奢华住宅的形式窠臼，探索中国高端居住模式

玻璃幕墙结合拱形金属屋面的设计为使用者带来从外至内的独特体验；
透明性材质消弭了室内空间与外部景观的隔阂；
玻璃映射让整体建筑在景色中呈现轻盈之姿；
大胆引入曲线元素，塑造出建筑内外空间的艺术感染力；
弧形穹顶高耸入云，为室内带来高敞体验；
屋面截面的螺旋曲线选用钢结构金属屋面系统呈现完美的曲线效果；
经过专业人士的多次测量计算，最后选用多种手段满足住宅的保温、防水隔汽、隔音需求；
整个屋面厚度为630 mm（不含吊顶厚度），共15个构造层次。

· Breaking Through The Stereotypes Of The Chinese Luxury Residences, Exploring The High-End Residential Model In China

The combination of the glass curtain wall with the arched metal roof brings users a unique experience outside to inside.
The transparent material eliminates the barrier between the interior space and the exterior landscape.
The reflection from the glass makes the whole building look lighter in the landscape.
Unapologetically introduce the curved elements to shape the artistic appeal of the interior and exterior spaces of the building.
The curved dome soars into the sky to make the interior wide and open.
The spiral curve of the roof section uses the steel-structure-metal-roof system to present a perfect curve effect.
After many measurements and calculations made by the professionals, a variety of methods were finally selected to meet the residential needs for the heat preservation, water and steam insulation and sound insulation. The thickness of the entire roof is 630mm (excluding the suspended ceiling), with a total of 15 structural levels.

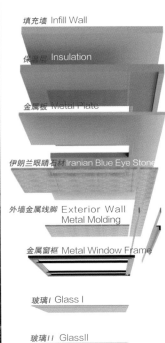

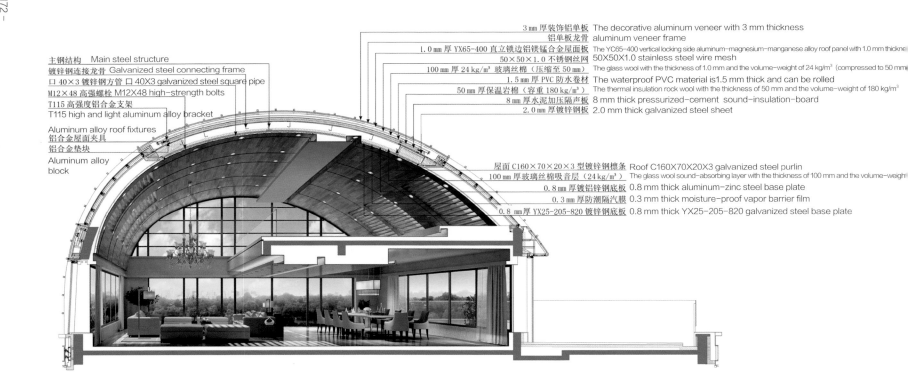

主钢结构 Main steel structure		3 mm 厚装饰铝单板 The decorative aluminum veneer with 3 mm thickness
镀锌钢连接龙骨 Galvanized steel connecting frame		铝单板龙骨 aluminum veneer frame
口 40×3 镀锌钢方管 口 40X3 galvanized steel square pipe		1.0 mm 厚 YX65-400 直立锁边铝镁锰合金屋面板 The YC65-400 vertical locking side aluminum-magnesium-manganese alloy roof panel with 1.0 mm thickness
M12×48 高强螺栓 M12X48 high-strength bolts		50×50×1.0 不锈钢丝网 50X50X1.0 stainless steel wire mesh
T115 高强度铝合金支架 T115 high and light aluminum alloy bracket		100 mm 厚 24 kg/m³ 玻璃丝棉（压缩至 50 mm） The glass wool with the thickness of 1.0 mm and the volume-weight of 24 kg/m³ (compressed to 50 mm)
铝合金屋面夹具 Aluminum alloy roof fixtures		1.5 mm 厚 PVC 防水卷材 The waterproof PVC material is1.5 mm thick and can be rolled
铝合金垫块 Aluminum alloy block		50 mm 厚保温岩棉（容重 180 kg/m³） The thermal insulation rock wool with the thickness of 50 mm and the volume-weight of 180 kg/m³
		8 mm 厚水泥加压隔声板 8 mm thick pressurized-cement sound-insulation-board
		2.0 mm 厚镀锌钢板 2.0 mm thick galvanized steel sheet
		屋面 C160×70×20×3 型镀锌钢檩条 Roof C160X70X20X3 galvanized steel purlin
		100 mm 厚玻璃丝棉吸音层（24 kg/m³） The glass wool sound-absorbing layer with the thickness of 100 mm and the volume-weight of 24 kg/m³
		0.8 mm 厚镀锌钢底板 0.8 mm thick aluminum-zinc steel base plate
		0.3 mm 厚防潮隔汽膜 0.3 mm thick moisture-proof vapor barrier film
		0.8 mm 厚 YX25-205-820 镀锌钢底板 0.8 mm thick YX25-205-820 galvanized steel base plate

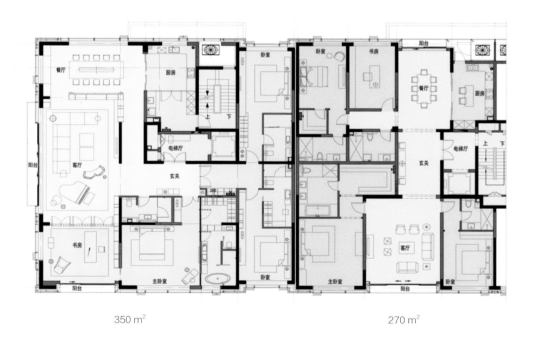

350 m²   270 m²

350+270 平面图
350+270 plan

重庆龙湖舜山府

CHONGQING LONGFOR SHUNSHAN MANSION

> 评委点评　JURY COMMENTS
>
> 总平面空间丰富，成功地创造了6个不同效果的庭院空间；立面简洁大气，富有时代感，很好地阐述了高端住宅的气质。
>
> The total plane space is rich, and the design successfully creates six different effects of the courtyard space. The facade is simple but elegant, full of the sense of the times, illustrating the traits of high-end residences.

第五届地产设计大奖·中国　居住项目－金奖
Residential Project | Gold Award of the 5th CREDAWARD

业主单位：重庆龙湖	Owner: Longfor Chongqing
规划设计：成执设计	Planning Design: Challenge Design
建筑设计：成执设计	Architectural Design: Challenge Design
景观设计：SWA	Landscape Design: SWA
道远设计	DAOYUAN Design
幕墙设计：上海安岩设计咨询有限公司	Facade: Anyan Design
项目地点：中国·重庆	Location: Chongqing, China
项目规模：79 287 m²	Project Scale: 79,287 m²
建筑面积：283 660.66 m²	Floor Area: 283,660.66 m²
建成时间：2018年	Date of Completion: 2018
设计周期：29个月	Cycle of Design: 29 months

项目用地处于重庆市北部新区照母山片区，有较成熟的城市交通体系。方案延续照母山肌理并加以梳理，层层递进，通过一环、一带、多轴线与照母山紧密连接，环山抱水，美不胜收，形成独一无二的高价值都市场所。项目承续中华文化，秉承传统园林尊贵居住风尚，园与府完美结合，巨献风雅礼序空间。山水之城以流线型的极具现代感的建筑形态与照母山的山形水势相呼应、相观望；强调立面的水平感和流动感，建筑风格极具未来感和标识性，创造高品质景观社区与自然环境的和谐对话。

With a developed urban transportation system, the project is located in the Zhaomu Mountain Area, in Chongqing. Combined with the mountain texture to go further layer by layer, the project is surrounded by water and mountains and closely linked with the Zhaomu Mountain through one ring, one belt, and the multi-line axis, forming a unique high-value urban place. Inheriting Chinese culture, adhering to the noble living style of traditional gardens, the project makes the garden perfectly melt into the mansion, presenting an elegant and ritual space. In addition, the project echoes the shape of the mountain with a streamlined and modern architectural form to emphasize the sense of level and fluidity of the facade. With the futuristic and iconic style, the project keeps a good balance between the high-quality landscape communities and nature.

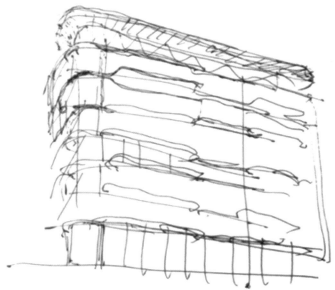

· 藏山纳水理念，融于弧形立面设计
建筑立面考究了山的光影变化，流线型的建筑形态带来一种"矛盾"的美，强调的是建筑与资源和生活的结合，形成了地景与人类活动相融合的奇妙场景，在一系列的事物所形成的氛围中创造出了可被感知的空间与建筑实体。

· Making The Existing Mountains And Water Melt In The Curved Facade Design
The facade of the building considers the change of the light and shadow of the mountain, and the streamlined architectural form presents a kind of "contradictory" beauty. The design emphasizes the combination of the architecture, the resources and life, forming the marvelous scene of the coexisting of the landscapes and human activities, and creating the touchable space and the architectural entities in the atmosphere formed by a series of things.

君山生活美学馆
JUNSHAN CULTURAL CENTER

> **评委点评**　JURY COMMENTS
>
> 设计师将北京当地传统民居的建筑手法进行精炼提取，糅合现代的建筑语言对其进行重新诠释，让建筑融于自然，使人与自然纯粹地结合。
>
> The designers assimilated the architectural techniques of the local traditional residential buildings in Beijing and reinterpreted them in modern architectural language. The project is to let the architecture melt in nature and people get well along with nature.

第五届地产设计大奖·中国　居住项目－金奖
Residential Project | Gold Award of the 5th CREDAWARD

业主单位：阳光城北京区域公司	Owner: Yango Beijing
规划设计：北京墨臣工程咨询有限公司	Planning Design: MoChen Architects & Engineers
建筑设计：如恩设计研究室	Architectural Design: Neri&Hu Design and Research Office
景观设计：广州山水比德设计有限公司北京分公司	Landscape Design: S.P.I Landscape Group
室内设计：如恩设计研究室	Interior Design: Neri&Hu Design and Research Office
照明设计：优米照明设计	Lighting Design: Lumia Lighting Design
项目地点：中国·北京	Location: Beijing, China
项目规模：4 277 m²	Project Scale: 4,277 m²
建筑面积：4 277 m²	Floor Area: 4,277 m²
景观面积：7 230 m²	Landscape Area: 7,230 m²
室内面积：4 277 m²	Interior Area: 4,277 m²
建成时间：2018 年	Date of Completion: 2018

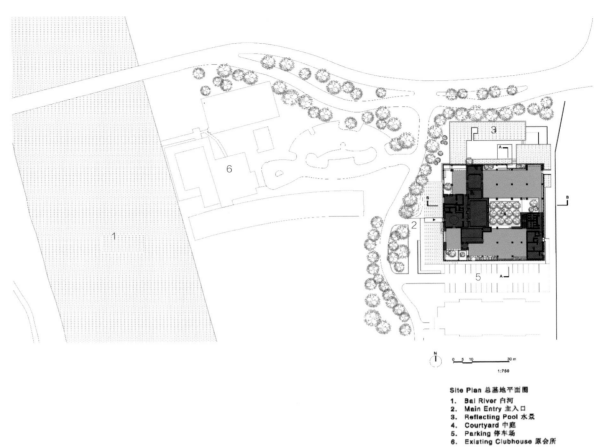

Site Plan 总基地平面图
1. Bai River 白河
2. Main Entry 主入口
3. Reflecting Pool 水泉
4. Courtyard 中庭
5. Parking 停车场
6. Existing Clubhouse 原会所

君山生活美学馆坐落于北京市郊群山起伏、河流蜿蜒的密云水库附近。建筑师从建筑本身的历史脉络中汲取灵感，将北方传统建筑与当代建筑语言相结合，从而转化为一种新的建筑表达。雕塑式的建筑形体有如从水面升起，与内外花园相连接，模糊了建筑和周围自然环境的边界。

Junshan Cultural Center is near the Miyun Resevoir, in Beijing, which is surrounded by the undulating mountains and winding rivers. Inspired by the historical context of the building itself, the architecture combines traditional northern architecture with contemporary one to express the new building, that is, the Junshan Cultural Center, which is like a sculpture, rising out of the water and linking the inside garden and outside one to make the building and its surroundings boundaryless.

· 立面上的形态呼应
立面上覆盖着一层轻盈的竹木格栅，内部空间偶尔会穿透建筑体量，在格栅上形成不同形状和深度的开洞，让各个立面在统一效果的同时也以不同的形态倒影呼应周边环境细节。

· Echoing Form On The Facade
The facade is covered with a layer of the light bamboo and wood grilles in which the openings with the different shapes and depths are formed due to the internal space occasionally penetrating the volume of the building, thus allowing each facade to unify the effect and reflect the details of the surroundings in different forms.

整座建筑由回收青砖砌成，展现浑厚的工艺内涵和历史感。铺地主材料选用的御窑金砖与青砖墙一同刻画主要的功能空间，使室外向室内的空间分布自然地过渡。

The whole building is made of recycled blue bricks, showing the profound technical connotation. The bricks of imperial kiln used for the main material for the paving, together with the blue brick walls, depict the main functional space, naturally connecting the outdoor with the indoor.

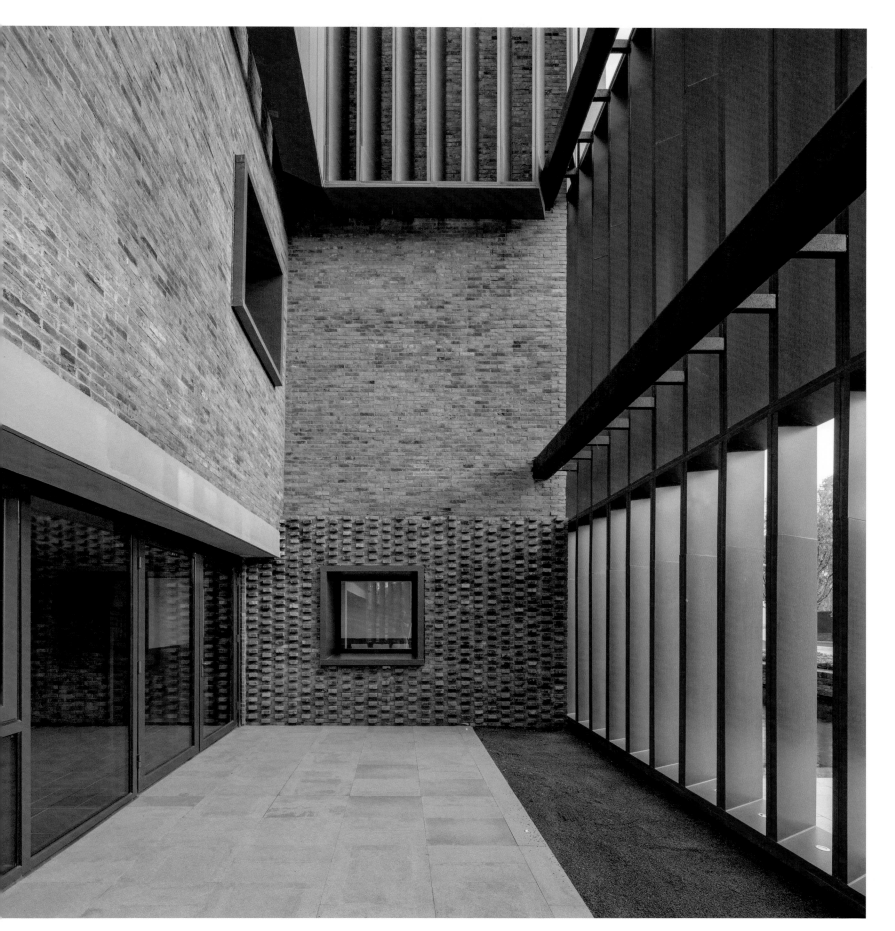

一系列定制家具和精致灯具的运用，搭配柔软的布料，再加上黄铜和自然石材的点缀，营造出低调而奢华的空间氛围。

The layering of customized furniture, refined brass metal detail, natural veins of stone accents, softness of fabric, and delicate lighting elements work together to compose a sense of understated luxury.

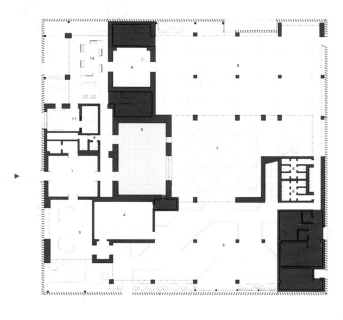

1F Plan 首层平面图

1. Lobby 大堂
2. Multi-Purpose Hall 多功能厅
3. Sales Reception 销售接待厅
4. Media Room 影音室
5. Winter Garden 四季花园
6. Restrooms 卫生间
7. Courtyard 中庭庭园
8. Art Exhibition Hall 艺术展厅
9. Bar 酒吧
10. Executive Lounge 商务休息室
11. Private Clinic 名人诊所

瀚海晴宇
HANHAI LUXURY CONDOMINIUMS

评委点评　JURY COMMENTS

本项目以创新设计为核心动力，颠覆市场对传统住宅项目认知的诉求，成为区域豪宅地产的标杆。

With an innovative design as the core driving force, this project updates people's cognition of traditional residential projects, making it the benchmark for the luxury home in that region.

第四届地产设计大奖·中国　居住项目－银奖
Residential Project | Silver Award of the 4th *CREDAWARD*

业主单位：瀚海大观地产	Owner: HANHAI DAGUAN REAL ESTATE
规划设计：双栖弧建筑事务所 　　　　　筑弧建筑事务所	Planning Design: amphibianArc 　　　　　　　　　archimorphic
建筑设计：双栖弧建筑事务所 　　　　　筑弧建筑事务所	Architectural Design: amphibianArc 　　　　　　　　　　　archimorphic
景观设计：AECOM	Landscape Design: AECOM
室内设计：弘品设计	Interior Design: Hongpin Design
幕墙设计：中南幕墙	Facade: Zhongnan Curtain Wall
摄　　影：张虔希	Photographer: Terrence Zhang
项目地点：中国·郑州	Location: Zhengzhou, China
项目规模：65 333 m²	Project Scale: 65,333 m²
建筑面积：225 000 m²	Floor Area: 225,000 m²
建成时间：2017 年	Date of Completion: 2017
设计周期：17 个月	Cycle of Design: 17 months

项目位于郑州市郑东新区，总建筑面积约 260 000 m²，容积率 3.5，包括 2 层地下车库、9 栋住宅塔楼、1 个豪华会所、一个 9 个班的幼儿园，以及独立的社区配套设施。项目的户型面积分别是：90 m²、180 m²、220 m²、330 m² 和 650 m²。会所包含了咖啡厅、茶室、健身房、SPA、游泳池及商铺，为业主提供了便捷的服务设施。

Located in Zhengdong New District, Zhengzhou City, the residential area named after Hanhai Luxury Condominiums, with 260,000 m² construction area and the Floor Area Ratio (FAR) of 3.99, covers one 2-story underground garage, nine residential towers, one luxurious clubhouse, one kindergarten with nine classes, and the independent community facilities. The floor area of the owners' home is 90 m², 180 m², 220 m², 330 m² and 650 m². The multi-functional clubhouse offers the owners various convenient services, such as the coffee shop, tea room, gym, SPA, swimming pool and shops.

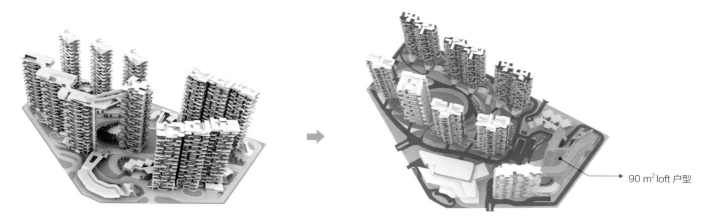

调整前布局 Before Adjusting　　　→　　　调整后布局 After Adjusting

90 m² loft 户型

· 户外生活高层化
可移动幕墙与户外设计相结合，将室外景观带入室内，在城市竖向居住空间内也能体验到类别墅的优越感和舒适感。

· 住宅设计的公建化处理
主楼立面设计的深色铝板及玻璃幕墙板与白色的流线型露台、阳台形成强烈对比，塔楼的跳层处理和公共建筑有机、流线的外形更凸显出整体建筑弧线设计的轻盈灵动。

· 稀缺的大平层独墅楼王——空中别墅，隐居于市
650 平方米大平层优美独特的立面使其不仅成为了晴宇项目的地标建筑，甚至已经成为了开发商的企业形象标志。

· 项目对城市展示面
在临主干道的基地边缘放置了各自不同的体量和具有特色的建筑形制，使得整体建筑群极具昭示性，对城市建设进行多方位的延展。

· 高容积率 v.s. 豪宅
塔楼沿基地边缘布置满足 3.5 的项目容积率要求，超大面积的中央绿化景观面积交错就高端社区的中央景观要素。

· 小户型政策 v.s. 豪宅
小户型集中规划，群居体量呈阶梯形状层层跌落，形成了趣味性和识别性兼得的地标式建筑。

· High-Rise Outdoor Activity
Combining the movable curtain wall with the outdoor design brings the outdoor landscape into the interior, making the owners staying at the vertical urban living place taste the feeling as good as that of living villas.

· Treating Home Facade Like Treating Public Building Facade
The dark aluminum panels and the panels of the glass curtain wall of the facade of the main building are in sharp contrast with the white streamlined terraces and balconies. In addition, the jump-story treatment of the residential tower and the public buildings featured with the streamlined shape highlights the lightness of the arc-shape building.

· The Precious And Rare Detached Villa With Penthouse——Sky Villa, Living Secludedly In The City
With the beautiful and unique facade and the 650 m² penthouse, the villa becomes not only the landmark building of Qingyu Project but also the symbolic image of the company.

· Its Impact On The City
Erecting at the edge of the site adjacent to the main road, the buildings, with different sizes and distinctive architectural shapes, are very attractive, making the city extend in multiple directions.

· Big Far V.s. Luxury House
The layout of the residential towers along the edge of the site meets the demand for the FAR of 3.5, and the super-large central green landscape area meets the demand for the element of the central landscape of the high-end community.

· The Policy Of Small Home V.s. Luxury Home
With the small home centralized planning, the size of the group living decreases layer by layer in a stepped shape, forming a landmark building with both fun and identification.

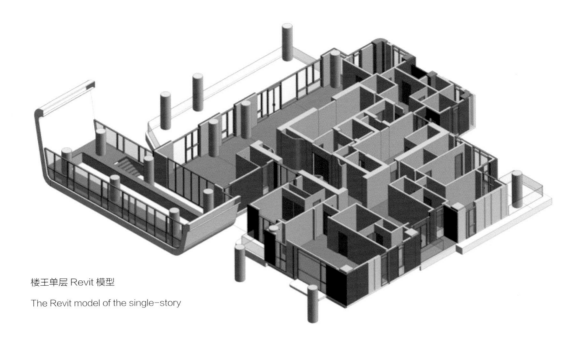

楼王单层 Revit 模型

The Revit model of the single-story

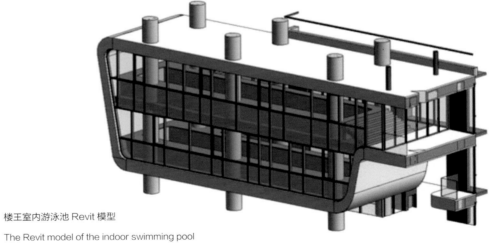

楼王室内游泳池 Revit 模型

The Revit model of the indoor swimming pool

· **设计和施工中对 BIM 模型的应用**

设计过程中应用 Revit 对建筑进行三维建模及出图，并且运用三维 Revit 模型直接和幕墙公司沟通，帮助合伙伙伴更好地理解设计，完美地保证了复杂形体的施工完成度。

· **Application Of Bim Model In Design And Construction**

Revit was used in the design to model and produce 3D drawings. In addition, the 3D Revit model was used to directly communicate with the curtain wall company to help its partners better understand the design to ensure the completion of complex shapes.

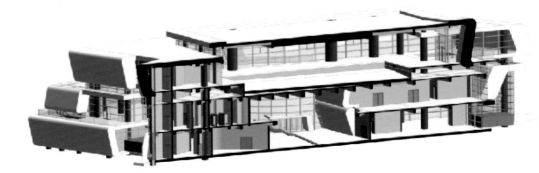

会所 Revit 模型

The Revit model of the clubhouse

创新设计和经济效益的完美结合造就豪宅标杆项目

The perfect combination of the innovative design and the economic benefits creates a benchmark project for luxury residences.

上海仁恒公园世纪
SHANGHAI YANLORD PARK CENTURY

评委点评　JURY COMMENTS

本项目通过建筑创意与技术的融合，创造前瞻性、国际化、地标性的高尚住宅社区，完美打造了仁恒第五代国际社区，公园世纪成为新一代高端住宅标杆项目。

Thanks to the architectural creativity and technology, the fifth generation of International Community of Yanlord-The Park, featured with advance, internationality and landmark, built a high-end community for its residents, making it become the benchmark project for the high-end residential area in a new generation.

第五届地产设计大奖·中国　居住项目－银奖
Residential Project | Silver Award of the 5th *CREDAWARD*

业主单位：上海仁恒置地	Owner: Yanlord Land
规划设计：杰作建筑设计咨询（上海）有限公司	Planning Design: TAL the architects. Co., Ltd.
建筑设计：杰作建筑设计咨询（上海）有限公司	Architectural Design: TAL the architects. Co., Ltd.
景观设计：NBBJ	Landscape Design: NBBJ
易亚源境	YAS Design
TIANHUA 天华	TIANHUA
中亚园林	Zhongya Landscape
项目地点：中国·上海	Location: Shanghai, China
项目规模：55 775.7 m²	Project Scale: 55,775.7 m²
建筑面积：210 263.85 m²	Floor Area: 210,263.85 m²
建成时间：2016 年	Date of Completion: 2016
设计周期：8 个月	Cycle of Design: 8 months

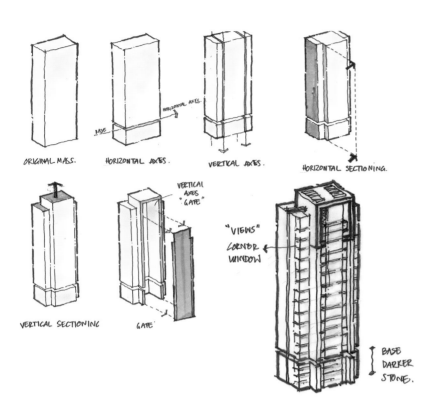

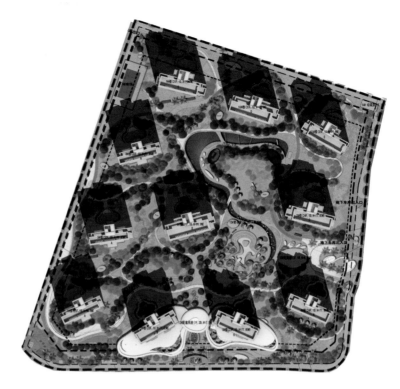

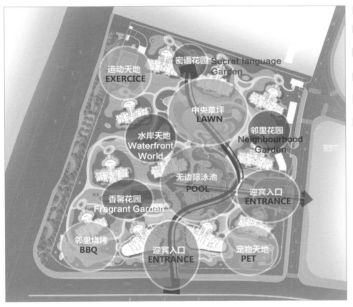

《诗经》云：春日迟迟，卉木萋萋。花木社区是上海滩四大花园之一"凌家花园"的发源地。本案位于花木社区的中心地块，毗邻世纪公园，自然环境绝佳。在设计构思中，突出营造富有活力的生态住区，将社区内部绿化景观向城市辐射，以此来创造前瞻性、国际化、地标性的社区环境，以至高生活享受之道为标准，致力打造出一个生态、健康、景观相结合的"都市高尚国际住宅区"。

The area of Huamu Community is the birthplace of "Lingjia Garden", one of the four major gardens in Shanghai, blessed with the beauty like *the Book of Songs* goes: "Springtime is warm and sunny, and days drag on, the grass and the trees are growing luxuriantly". Located at the central area of Huamu Community, Yanlord-The Park, which is adjacent to the Century Park, is surrounded by the beautiful natural environment. Its design aims to create a vibrant eco-friendly residential area by means of making the inside green landscape of the community radiate to the city to build a community of being forward-looking, international and land-marking. Taking enjoying the quality life as the standard, its design strives to set an example of " The Urban Distinguished International Residential Area" featured with ecology, health and views.

· 立体景观住宅

景观与观景并存、经典与创新交融，秉承"立体景观住宅景观理念"引入生态会所和阳光地库，筑成一方具有活力和生态舒适的生活大环境。

· 现代简约主义风格

立面整体风格简洁、挺拔，转角窗设计及 350mm 低窗台设计让业主深切享受到最佳角度的公园景观。

· 3D Landscape Residence

With the coexisting of landscapes and views as well as the combination of classics and innovations, the design, adhering to "the residential landscape concept featured with three-dimensional landscape", strives to build a vibrant living environment of being eco-friendly and comfortable by means of using the ecological club and sunny basement.

· The Style Of Being Modern And Concise.

The overall style of the facade is concise and straight. In addition, the design of the corner window and the 350mm low window makes the owner enjoy the best view of the park.

· **立面排版 | 错落有致**
铝板排版打破横平竖直的样式，追求错落有致、不落俗套的设计效果，节约用料、美观大方。

· **Layout: Being Well-Arranged**
The layout of the aluminum plates breaks the style of being horizontal and vertical, pursuing the features of being unique, well-arranged, economic and beautiful.

· 双大堂入户体验
豪华地下大堂不占用容积率面积，有效提高入住体验，独特设计的唯一性提高业主极致感。

· 阳光地库
充分结合双大堂优势，车道中间结合地面景观设有玻璃采光天窗，将阳光与自然直接引入地下。

· The Double-Lobby Experience
The luxurious underground lobby consumes none of the area of the floor area ratio to please the owners.

· Sunny Basement
Make best of the double-lobby to use the glass skylight to directly bring the sunlight and nature into the underground.

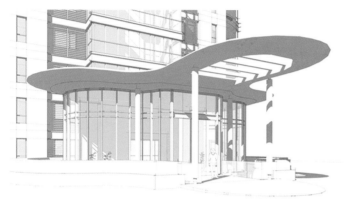 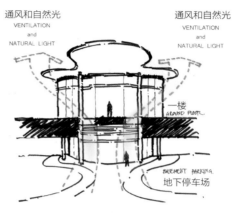 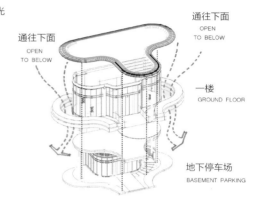

03 景观设计 LANDSCAPE DESIGN

01 黄山雨润涵月楼 —————————————————————————— 208
HUANGSHAN YURUN HANYUELOU VILLA RESORT

02 上海万科翡翠滨江 ————————————————————————— 214
SHANGHAI VANKE EMERALD RIVERSIDE

03 福州万科城 ———————————————————————————— 222
FUZHOU VANKE CITY

228	莫奈大道 2.0 — MONET AVENUE 2.0	**04**
236	音昱水中天 — SANGHA BY OCTAVE	**05**
244	杭州萧山 前湾国际社区·无界公园 — XIAOSHAN HANGZHOU BAY UNBOUNDED PARK	**06**

黄山雨润涵月楼
HUANGSHAN YURUN HANYUELOU VILLA RESORT

> **评委点评　JURY COMMENTS**
>
> 景观设计注重本土特色语言，运用徽派建筑元素，尊重文化根源，巧用当地地形生态环境而成。使用土生植物及建材资源，以移步换景的手法打造养生休闲度假村。
>
> With Huizhou architectural elements, the landscape design, focusing on the native nature and respecting the local culture, skillfully uses the local terrain and ecological environment(such as, the local plants)and building materials to build a leisure resort in a way of changing scenery.

第一届地产设计大奖·中国　景观设计 – 金奖
Landscape Design | Gold Award of the 1st *CRED*AWARD

业主单位：黄山松柏高尔夫乡村俱乐部有限公司	Owner: Huangshan Pine Golf & Country Club Co., Ltd.
景观设计：贝尔高林国际（香港）有限公司	Landscape Design: Belt Collins International (HK) Limited
项目地点：中国·安徽·黄山	Location: Huangshan, Anhui, China
项目规模：165 000 m²	Project Scale: 165,000 m²
景观面积：165 000 m²	Landscape Area: 165,000 m²
建成时间：2013 年	Date of Completion: 2013
设计周期：69 个月	Cycle of Design: 69 months

黄山雨润涵月楼酒店采用传统徽南文化中建筑与景观的造型设计风格，完全以别墅作为独享空间。以完美服务、荟萃精英为愿景，酒店中的每栋别墅都将尊贵与私密、便捷与专业的服务作为运营宗旨，使每一位入住的客人能够充分感受家的温馨，以及比家更为畅然舒适的周到服务，犹如置身于优雅江南情境。

Provided with the private villas, the Hanyuelou Villa Resort, Huangshan is featured with the traditional southern Anhui architectural and landscape style. With the vision of offering perfect services and a place for elites gathering, the Hanyuelou Villa Resort, Huangshan, guided by the running principle that offers the distinguished, private, convenient and professional services, makes every resident fully feel as warm as staying at home and even better than at home to enjoy the elegant views of the region of the Yangtze Delta.

- 创新性的方案将皖南村落的景观特征渗透到徽派建筑当中，以原生态的设计方式让涵月楼山谷酒店融于皖南山地中。

本项目在原有徽派建筑风格的基础上采用了简约、清新的现代景观设计手法，创新性地将皖南村落的著名景观特色运用在本项目中——用现代镜面水池的方式将黄山脚下宏村的"半月塘"呈现在各个主要入口；棠樾"牌坊群"的门楼概念亦出现在几个主要入口及景观节点上。

- The creative and original idea melts the landscape features of the villages in southern Anhui melt into the Hui-style architecture. In addition, the original eco-design makes the Hanyuelou Villa Resort, in the mountains in southern Anhui.

Based on the existing Hui-style architecture, the project, with the concise and fresh modern landscape design technique, creates the original idea to absorb the famous landscape features of the villages in southern Anhui——the "Half-moon-shaped Pond" which locates at the Hong village at the foot of the Yellow Mountain. They are at the each main entrance with the technique of the reflecting pool. The concept of the "Memorial Archway Groups" of Tangyue village also appears at several main entrances and the landscape nodes.

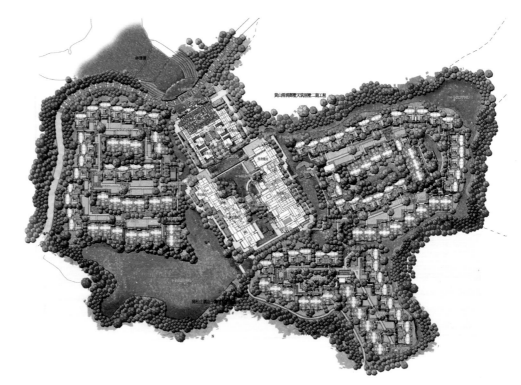

上海万科翡翠滨江

SHANGHAI VANKE EMERALD RIVERSIDE

> 评委点评　JURY COMMENTS
>
> 上海黄浦江边的万科翡翠滨江紧靠民生码头和老粮仓，景观设计希望对建筑场地赋予灵气和生机。梭形的文化中心建筑咬合东、西椭圆场地，以水为图底，步行小道、植栽池、休闲场地如行云流水般展开，落笔大气，收放有序，细节考究，实现度较高。
>
> The aim of the landscape design is to energize the Vanke Emerald Riverside by the Huangpu River in Shanghai, which is close to the Minsheng Wharf and the old granary. The spindle-shaped cultural center building adjoins the eastern and western elliptical sites. With water as the base, the walking trails, planting ponds, and leisure places are laid out delicately.

第一届地产设计大奖·中国　景观设计 – 金奖
Landscape Design | Gold Award of the 1st *CREDAWARD*

业主单位：万科集团	Owner: VANKE
景观设计：CICADA　朗道	Landscape Design: CICADA　LANDAU
项目地点：中国·上海	Location: Shanghai, China
项目规模：60,000 m²	Project Scale: 60,000 m²
建筑面积：128 910 m²(住宅)；47 256 m²(商业)	Floor Area: 128,910 m² (Residential Area); 47,256 m² (Commercial Area)
景观面积：45 969 m²	Landscape Area: 45,969 m²
建成时间：2014 年	Date of Completion: 2014
设计周期：18 个月	Cycle of Design: 18 months

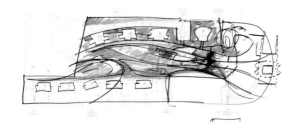

· **历史如长河一样在景观上流淌**

一条历史的长河将景观的轴线贯穿。上海万科翡翠滨江两个地块在规划上采用形式感各异的围合空间,经典与时尚碰撞的立面风格还原了城市生长的痕迹,呈现生动的滨江形象。

· **History liking a long river flows on the landscape.**

A historical river runs through the axis of the landscape. The two plots of SHANGHAI VANKE EMERALD RIVERSIDE are planned to use enclosed spaces with different forms. The facade style featured with classic and fashion collision restores the traces of urban growth and presents a dynamic image of the riverside.

上海万科翡翠滨江是目前上海内环内整个陆家嘴滨江沿线最大的城市综合体项目。在景观规划方面,本项目为致力于打造滨江沿线最佳高端社区,建造了一个似淮海公园大小的绿化景观。其作为浦江两岸综合开发的四个重点区域之一,未来将与老外滩、陆家嘴的顶级住宅区不分伯仲,构成上海中央腹地的"黄金三角"。

项目临近民生路码头,将斥资 10 亿元全面改造这个具有传奇色彩的码头。设计灵感也从这里获得,并将传奇延续下去。

As the largest urban complex area in Lujiazui by the Huangpu River within the inner ring of Shanghai, the green landscape of the complex area of the SHANGHAI VANKE EMERALD RIVERSIDE has the size as big as that of the Huaihai Park to build the best high-end community along the Huangpu river. As one of the four key areas for comprehensive development on both sides of the Huangpu river, the complex area of the Emerald Riverside will compete with the other two top residential areas respectively located at the Old Bund and Lujiazui in the future, forming the "Golden Triangle" in the central hinterland of Shanghai.

The project is adjacent to the legendary wharf completely renovated at a cost of one billion RMB and located at Minsheng Road. The larger-than-life wharf inspires the designer to design this project and keep its legend.

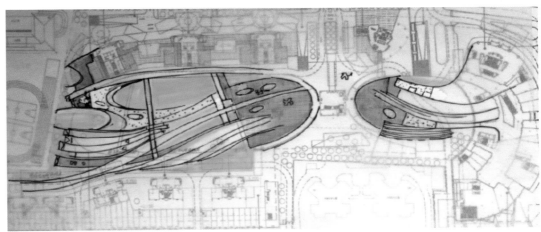

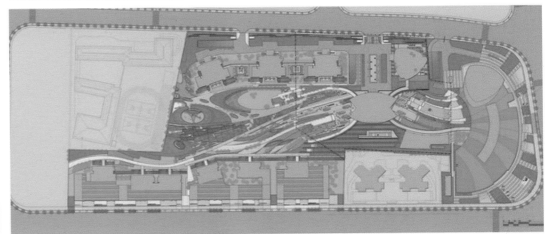

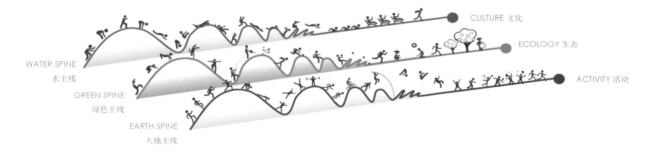

· 时间的禅意

三条设计轴线——"水主线""绿色主线""大地主线"由西向东交汇，演绎了具有时间特征的景观轴线。

"传统"——传统园林特色与现代工业清水混凝土的汇聚显现出自然融洽的向心性文化。

"现代"——开阔的广场和草坪体现多元化的感官形态，一路向西的铺装形式和水景也代表着岁月的流淌，共同汇聚成景观焦点。

"未来"——商业景观线条硬朗，大乔木与整形灌木、草坪进行组合，开放性的水景也预示着绿色环保的未来发展。

· 十里洋场中的禅

项目从时间、空间等多方位角度探寻万科陆家嘴翡翠滨江在传统、现代和未来之间碰撞出的清空与安宁，让设计生活在景观与禅意间尽显艺术美感。

· The Zen Of Time

Three design axes—— "Main Line of Water", "Main Line of Green" and "Main Line of Earth" converge from the west to the east, illustrating a landscape axis with time characteristics.

The element of "Tradition" —— the convergence of the traditional garden features and the modern industrial fair-faced concrete shows a natural and harmonious centripetal culture.

The element of "Modernity" —— the spacious square and greenland display the various forms of senses, the paving forms and the waterscapes along all the way to the west also symbolize that time goes like water. All in all, they become the spotlight of the landscape.

The element of "Future" ——The commercial landscape has strong lines. Big trees trimming shrubs and lawns are combined. As well as the open waterscape indicates the eco-friendly development in the future.

· The Zen Of The Glamorous Shanghai

With the sense of clearness and tranquility provided by the Vanke Emerald Riverside, its design, from the perspective of time and space, melts the tradition, the modernity into the future, showing the artistic beauty between the landscape and Zen.

· 空间的禅意
隐秘的瑜伽园和欢快的儿童活动园截然不同的矛盾空间在此处协调统一,动静皆宜,别具匠心。

· 细节的禅心
景观源头从西向东、从高向低延伸,地面与水池一齐从地坪升起形成坐凳和花池。水景和草地上的蛋形雕塑也保持着翻滚的形态,变化的细节寓意着时间的不断前行。

· The Zen Of Space
With the unique characteristics of the pleasant association of motion and stillness, the two contradictory places of the hidden Yoga Garden and the cheerful Children Amusement Park are coordinated and unified.

· The Zen Of The Details
The origin of the landscape extends from the west to the east, as well as, from the high to the low, decorated with benches and the flower pond. In addition, the egg-shaped sculptures on the waterscape and the lawns keep rolling. All those changes in details symbolize flying time.

福州万科城
FUZHOU VANKE CITY

评委点评　JURY COMMENTS

"桥·连接"的主旋律带动起了山谷、湖水、树木等自然景观的合唱，音符的变化和跳跃也使建筑有了生机，人也顺理成章地回到大自然的家。人化自然与原生自然的有机融合是本项目景观设计的最大特色。

The theme of the design is "Bridge · Connection" which leads valleys, lakes, trees, and so on to chorus. The changing and jumping notes gives life to the buildings and makes the residents return to the nature. The most attractive feature of this design is to make the humanized nature to melt into the native one perfectly.

第二届地产设计大奖·中国　景观设计 – 金奖
Landscape Design | Gold Award of the 2nd *CREDAWARD*

业主单位：福州市万科房地产有限公司	Owner: Fuzhou Vanke Real Estate Co., Ltd.
规划设计：SWA 洛杉矶办公室	Planning Design: SWA Los Angeles
景观设计：SWA 洛杉矶办公室	Landscape Design: SWA Los Angeles
摄　　影：SWA David Lloyd	Photographer: SWA David Lloyd
项目地点：中国·福建·福州	Location: Fuzhou, Fujian, China
项目规模：450 000 m²	Project Scale: 450,000 m²
建筑面积：800 000 m²	Floor Area: 800,000 m²
景观面积：350 000 m²	Landscape Area: 350,000 m²
建成时间：2015 年	Date of Completion: 2015
设计周期：12 个月	Cycle of Design: 12 months

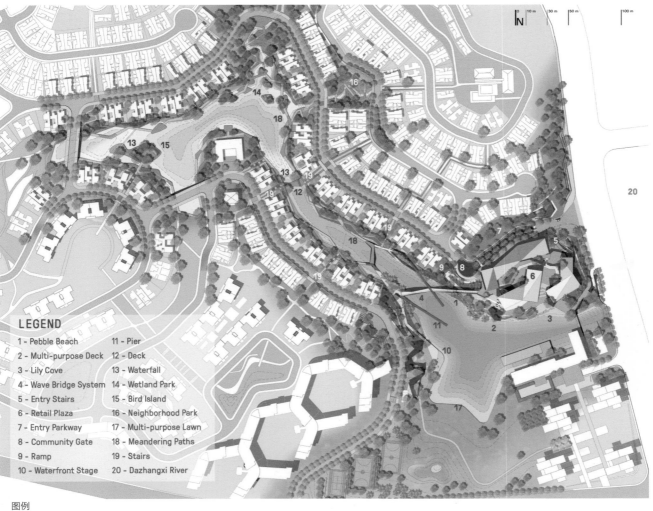

LEGEND
1 - Pebble Beach
2 - Multi-purpose Deck
3 - Lily Cove
4 - Wave Bridge System
5 - Entry Stairs
6 - Retail Plaza
7 - Entry Parkway
8 - Community Gate
9 - Ramp
10 - Waterfront Stage
11 - Pier
12 - Deck
13 - Waterfall
14 - Wetland Park
15 - Bird Island
16 - Neighborhood Park
17 - Multi-purpose Lawn
18 - Meandering Paths
19 - Stairs
20 - Dazhangxi River

图例
1. 卵石滩
2. 多功能甲板
3. 百合湾
4. 波浪桥系统
5. 入口楼梯
6. 零售广场
7. 公园大道入口
8. 社区门
9. 斜坡
10. 海滨舞台
11. 码头
12. 甲板
13. 瀑布
14. 湿地公园
15. 鸟岛
16. 街区公园
17. 多功能草坪
18. 漫步小路
19. 楼梯
20. 大漳溪

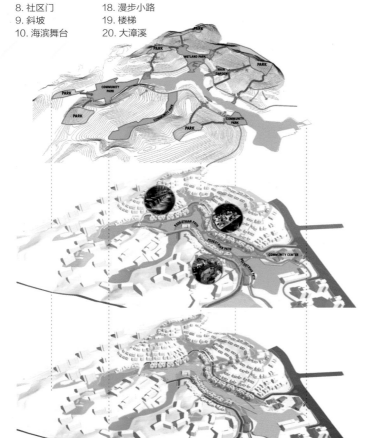

○ Park & Greenway System (From Mountain To Valley) 公园和绿道系统（从山到谷）
○ Stormwater Collects Follow The Greenway System (From Mountain To Valley)
沿着绿道系统的雨水收集系统（从山到谷）

○ Pedestrian Path Meandering Along Lakeside 湖边漫步的人行道
○ Bridging System Provides Linkage & Gathering Space
桥接系统提供联动和聚集空间

○ Residential Community Settles At The Valley (Great Land Value) 位于山谷的住宅社区（土地价值巨大）
○ Planning Of Public Space Is Based On The Existing Terrain (Mountain Peak & Waterfront)
基于现有地形的公共空间的规划（山峰和海滨）

项目整体规划以具有自然地势的山和谷作为住宅社区组团的分区；利用现有溪谷地形，以生态手法创建了一个如画般的核心湖区作为贯穿整个基地的公共开放空间。主要设计理念——"缝织"的景观——除了在整体规划上透过地势及水系的条件组成交织的公共空间网络，在实质的空间层面，步道、桥、廊、水景等也都以转折交错的形式呈现出空间设计的独特性。

福州万科城是一个住宅和综合用途开发项目，开发计划新建精致酒店、商店街、会所、集合住宅、连排和独栋别墅等等不同类型的住宅。基地东侧临大樟溪，坐落在中国东南赤壁风景区的连绵起伏的丘陵上。占地45公顷的场地距离福州市中心不到30分钟车程，这使它成为越来越多的城市专业人士购买第二套房的热门地区。与许多工业化国家一样，现代化和城市化虽然带来了奢华的生活方式，但人们仍然渴忘着郊区的田原和自然风光。

Taking the natural mountains and valleys as the districts of the residential communities, the overall plan of the project based on the existing brook and valley creates a picturesque lake area in an eco-friendly way, which serves as the public open space throughout the whole site. The main design concept is to stitch and weave the landscapes, that is, generally, the intertwined public space network is built on the existing topography and water system; physically, the uniqueness of the space design is displayed in the form of turning and knitting the trails, bridges, corridors, waterscapes and so on.

Fuzhou Vanke City is a residential and mixed-use development nestled in the rolling hills of the Red Cliff Scenic Area in Southeast China. The development program includes a mix of a boutique hotel, a shopping street, clubhouses, residential high-rises, apartment buildings, townhouses and single-family houses. The 45-hectare site is within a 30-minute drive to Fuzhou, making it a popular area for the rising class of urban professionals to purchase a second home. Like many industrialized countries, although modernization and urbanization have introduced luxurious lifestyles into China, people still long for the countryside and natural scenery in the suburbs.

湖中的人行景观桥结构尤为重要。这座桥以一种有机的形式上下波动,同时还通过光滑的冲孔铝板立面展示了工程和材料方面的技术。就像传统的中国园林一样,桥架抬高了行人,使观者有更开阔的视觉体验。在最高处,桥下划艇还可通过,在水上创造出层次感。桥被设计为连接湖南北岸的社区,设计提供了一种完全不同的自然观,一种在湖上漂浮的感觉。在这个空间中,孩子们可以跳过父母的视野,欣赏湖边和远处群山的壮丽景色。桥的构造经过了多次研究,才能形成美观、明晰且具有成本效益的结构。设计团队通过人行桥和景观设施,使人们能更加接近自然、感受自然。

The structure of the pedestrian bridge on the lake is of particular importance. The bridge not only harmoniously fluctuates up and down, but also illustrates the technology of its engineering and materials with the smooth perforated aluminum facade. The bridge plays the same role as bridges in the traditional Chinese gardens do to raise its pedestrians and broaden their views. In addition, it allows rowing boats to pass through its top place, creating layers on the water. The design of this bridge aims to connect the northern communities on the bank of the lake with the southern ones, providing a feeling of floating for its pedestrians, and allowing kids to skip their parents' vision to enjoy the magnificent view of the lake and the distant mountains. Furthermore, forming a beautiful, clear and cost-effective bridge, its structure has been studied for many times. Obviously, the design team hopes to bring nature close to people through introducing the pedestrian bridge and landscape facilities into their design.

生态永续是这个项目的核心价值，从长期维护管理的角度来看，中央湖区需要达到自我水生态的稳定和生物种的平衡，人工湖水量需在全年的枯水期和丰水期达到平衡，水循环和排洪涝等设施经过专家的研究被纳入了景观设计之中。水中和水岸等湿生植物成为水生态景观的重要构成。大樟溪的原始自然生态良好，现有植被丰富，设计初期尽可能地保留现有的大树，包括马尾松、樟树、板栗、柞木和枫香等大乔木以及竹类等。在设计后期，项目将引入福建乡土植物，如盆架木、木棉、白兰、芒果、黄花槐等；还有果树，如荔枝、杨梅、青梅、枇杷等，还有丰富的花灌木，如黄蝉、野牡丹、美人蕉、狗牙花、山茶、栀子、南天竺、茉莉花等；还将增加具山野气息的各类观赏草，如狼尾草、芒类等，形成植被丰富、花色突出、季相变化明显、富有自然野趣的山水旅游景观。

The stability and sustainability of the ecological system are the core value of this project. From the perspective of long-term maintenance and management of this project, the design has to consider the following aspects: The central area of this lake must keep the balance between the stability of its own water ecology and its biological species; the water volume of this manmade lake must be balanced during the dry season and the flood one of the years; the experts' proposal is that the facilities of water circulation and flood discharge should be considered. In addition, the hygrophytes along the banks and in the water are the important elements of the water ecological landscape. Due to the good natural eco-system and the rich existing vegetation in Da Zhangxi, the design of this project proceeds in two phases: At the early stage, the existing big trees are retained as many as possible, such as Masson pine, Camphor tree, Castanea millesimal, Oak, and Sweetgum, and Bamboo; In the later stage, the various plants are introduced in this project, such as the native plants in Fujian will covering the potted wood, kapok, white orchid, mango, yellow locust, the fruit trees including lychee, bayberry, green plum, loquat, the rich flowering shrubs like yellow cicada, wild peony, canna, dogtooth flower, camellia, gardenia, nan Tianzhu, jasmine, and the variety of ornamental grasses like Pennisetum, mangosteen, thus forming an attractive place with the rich vegetation, characteristic flowers, clear four seasons, and natural wildness.

莫奈大道 2.0
MONET AVENUE 2.0

> **评委点评　JURY COMMENTS**
>
> 莫奈大道商业街区的改造设计提供了一个成功范例，让我们看到如何通过景观和街景设计来改善城市空间，在车行交通为主的城市创造宜人的场所。该景观设计不仅提升了周边的地产价值，更是通过设计来积极推动可持续发展的愿景，创造具有宜人尺度的环境，打造健康的都市生活方式。项目细部的完成度也可圈可点。
>
> How to improve the urban space through designing the landscape and streetscape to create a pleasant place in a heavy traffic city, the answer exists in the renovation of Monet Avenue, the commercial street. The landscape design not only enhances the value of the surrounding real estate, but also actively promotes the sustainable development, builds a pleasant environment, and creates a healthy urban lifestyle. In addition, the detailed design is remarkable.

第三届地产设计大奖 · 中国　景观设计 – 金奖
Landscape Design | Gold Award of the 3rd CREDAWARD

业主单位：森林城市商业集团 & 西海岸商业发展公司	Owner: Forest City Commerical Group & West Coast Commercial Development
景观设计：SWA	Landscape Design: SWA
项目地点：加利福尼亚州	Location: California
项目规模：9 288 m²	Project Scale: 9,288 m²
建筑面积：9 288 m²	Floor Area: 9,289 m²
景观面积：4 644 m²	Landscape Area: 4,644 m²
建成时间：2015 年	Date of Completion: 2015
设计周期：8 个月	Cycle of Design: 8 months

莫奈大道 2.0 项目给予了公众无法从电子商务中所获得的愉悦、令人回忆的购物体验。如今在实体店购物，人们更关注的是消遣体验而非便利性。我们所配建的独特景观建筑就是为了创造出这样的体验，特别是在室外零售环境方面。

莫奈大道 2.0 延续了维多利亚花园致力于提供高质量公共空间的努力外，还增加了艺术性和可持续性，同时，通过人行道和公共空间的建设为激烈竞争中的零售物业和商业街区提供了景观建筑解决方案。独特的景观建筑，塑造了高质量发展的艺术空间。

The Monet Avenue 2.0 provides the public with an enjoyable, memorable experience which the e-commerce cannot offer. Nowadays the people who go shopping at the physical stores prefer the recreational experience to convenience. To meet that goal, we create this unique landscape building to offer the outdoor retail environments for people to taste this type of experience.

The Monet Avenue 2.0 not only inherits the commitment of the Victoria Gardens to provide the quality public space, but also adds the additional layers of arts and sustainability. In addition, this project provides a landscape architectural solution for the struggling retail developments and main streets by implementing the sidewalks and the public spaces. Furthermore, equipped with the unique landscape architecture, the project builds a high-quality art space.

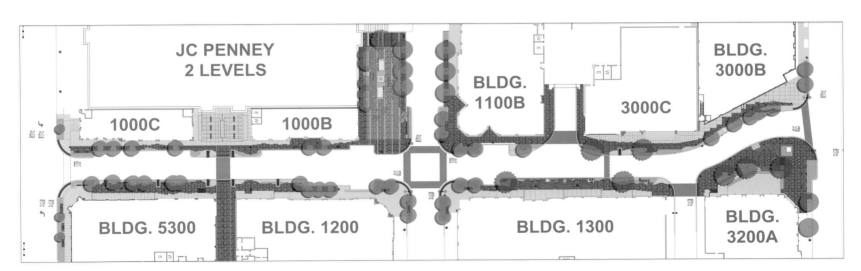

· 莫奈大道 2.0 致力于设计高质量的公共空间，增加艺术性的同时塑造绿色、可持续的景观特性。
· 通过人行道和公共空间的建设为激烈竞争中的零售物业和商业街区提供景观建筑的解决方案。

· The Monet Avenue 2.0 is committed to designing a high-quality public space, and increasing the artistic character, as well as, the green and sustainable landscape one.
· This project provides a landscape architectural solution for the struggling retail developments and main streets by implementing the sidewalks and the public spaces.

· 定制可回收木材长凳
· 石面上的激光切割图案来自圣盖博山脉的山脊线。

· Customized recyclable wooden benches
· The existing laser-cut pattern is derived from the ridgeline of the San Gabriel Mountains.

人们经常聚集在水景周围,中央喷泉使用了不断变化的喷泉水效果——雾和瀑布。

The scenic water attracts people to go there, the central fountain acts the changing water performance——fog and waterfalls.

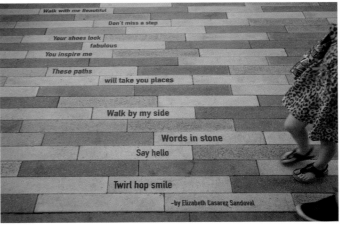

选择了当地高中生比赛中的一首诗嵌入铺路中。

Pave with an embedded poem selected from the local high school student contest.

音昱水中天
SANGHA BY OCTAVE

> **评委点评**　JURY COMMENTS
>
> 漂浮在阳澄湖上的低密度综合社区集度假生活、学习培训、医疗康养于一体，由四个浮萍式小岛构成。湖光、碧水、荷叶、绿植的"软"和褐色乱石砌筑的泊岸、糙石碎拼场地的"硬"结合得非常好。景观从整体到细节层层推进，很放松，很从容，营造了归隐自然、离世脱俗的湖区环境。
>
> Floating on the Yangcheng Lake, the low-density multi-functional community is composed of four duckweed-like islands, realizing various functions (such as vacation, study, and medical care) in this community. In addition, the spotlight of this design is that the "softness" elements (such as lake light, clear water, lotus leaves, and green plants) are perfectly combined with the "hardness" ones (such as the bank made of brown irregular stones, the site made of pieces of rough stones), making the community melt in the nature in a way of advancing the landscape layer by layer.

第五届地产设计大奖·中国　景观设计 – 金奖
Landscape Design | Gold Award of the 5th *CRED*AWARD

业主单位：苏州万邦置业发展有限公司	Owner: Suzhou Wanbang Real Estate Development Co., Ltd.
景观设计：地茂景观设计咨询（上海）有限公司	Landscape Design: Design Land Collaborative Ltd.
项目地点：中国·江苏·苏州	Location: Suzhou, Jiangsu, China
景观面积：40 826 m²	Landscape Area: 40,826 m²
建成时间：2017 年	Date of Completion: 2017
设计周期：34 个月	Cycle of Design: 34 months

"音昱水中天"的灵感来自梵文"社区"一词，它是一处位于苏州阳澄湖畔，集生活、工作、学习为一体的综合性社区，占地约19万平方米，毗邻拥有1 500年历史的重元寺。阳澄湖是连接太湖和长江水体的文化和生态纽带，客户的愿景是创设一个生活的范本，通过对活力文化的前瞻性研究，鼓励社会发展，激发21世纪的生活灵感。

该场地是一个废弃的西班牙风格的"商品住宅"项目，建在人工岛上，除非视觉影响、生态和场地与水的关系得到妥善处理，否则这些问题与客户的理念是对立的。景观设计采用当地的石头、砖和混凝土再生材料，营造了一个可再生的、可持续性和弹性的环境，"音昱水中天"已经成为一个有低层建筑的、优雅的美丽湖滨和季节性的本地花园。客户的愿景由环境设计实现，这有助于提醒游客，他们与自然并没有分离，而是联系在一起的，是自然世界的一部分。

Inspired by the Sanskrit word for 'community', Sangha is a 190 000 m² viable live-work-learn community on Yangcheng Lake near the 1,500-year-old Chongyuan Temple near Suzhou. Considered a cultural and ecological array of connecting water bodies between Lake Tai and the Yangtze River, Yangcheng Lake our client's vision was to create a model and inspiration for 21st-century living as a forward investigation on animating culture, encouraging social development and exploring diverse ways to engage its citizens in this dynamic country of China.

The site, a defunct Spanish style 'commodity housing' project built on an artificial island opposed our client's very ideology unless visual impact, ecology and the sites relationship with the water was properly addressed. Through a context-sensitive approach to renewable sustainability and resiliency, Sangha has become an elegant lakeside aesthetic of low-rise buildings and seasonal native gardens built with local stone and reclaimed materials of brick and concrete. Our client's vision, hand in hand with their newly adopted environment helps remind visitors that they are connected and part of the natural world rather than separate from it.

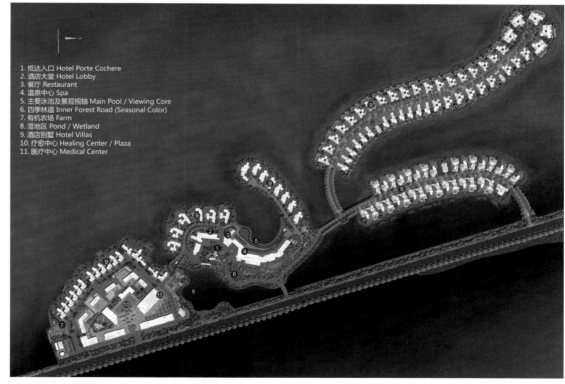

1. 抵达入口 Hotel Porte Cochere
2. 酒店大堂 Hotel Lobby
3. 餐厅 Restaurant
4. 温泉中心 Spa
5. 主要泳池及景观视轴 Main Pool / Viewing Core
6. 四季林道 Inner Forest Road (Seasonal Color)
7. 有机农场 Farm
8. 湿地区 Pond / Wetland
9. 酒店别墅 Hotel Villas
10. 疗愈中心 Healing Center / Plaza
11. 医疗中心 Medical Center

Spring 春

Summer 夏

Autumn 秋

Winter 冬

· **植栽策略**

与湖区生态相协调的植栽策略进一步提升了项目的品质。项目通过植栽的运用屏蔽视觉侵扰，创造私密区域，并敞开重要景观节点。富有层次的植栽营造季节性气氛，为全年四季增添乐趣，并提升湖滨的特殊品质。

· **Planting Strategy**

The project will be enhanced by a planting strategy that is in harmony with the ecology of the lake, screens visual intrusions, creates privacy where needed, and opens up important views. The layers of planting will create seasonal moods that will add to the year-round ambience and improve the special quality of the lakeside.

- 开花乔木 Flowering Tree
- 秋色森林 Fall Color Forest
- 规整式行道树 Formal Street Tree
- 自然式行道树 Natural Street Tree
- 水岸乔木 Water Edge Tree
- 泳池广场树 Main Pool Tree
- 乔木遮蔽 Tree Screen
- 混合乔木遮蔽及水岸乔木 Mix-tree Screen / Riparian Tree
- 常绿乔木遮蔽 Evergreen Tree Screen

景观设计师的第一种方法是通过创造"活力湖岸线"来缓解开发项目对水体的入侵。在某些情况下，场地现存的挡土墙被保留并进行了不同程度的消减，作为一种手段来最大程度地减少项目对自然环境可能带来的负面影响，并帮助弱化场地与水的生硬边界。这将有效地巩固设计策略，以在场地与水界面或沿岸区域内创造湿地，恢复活动和植物缓冲带。

The first approach as landscape architects was to soften the development's intrusion into the water by creating a 'living shoreline'. The existing coastal bulkheads from the previous development are preserved and manipulated, in some cases, as an effort to minimize the project's overall environmental impact, and to help assist blurring the land-water interface. This would efficiently sanction strategies to create wetland restoration efforts and vegetative buffers within the land-water interface or littoral zone.

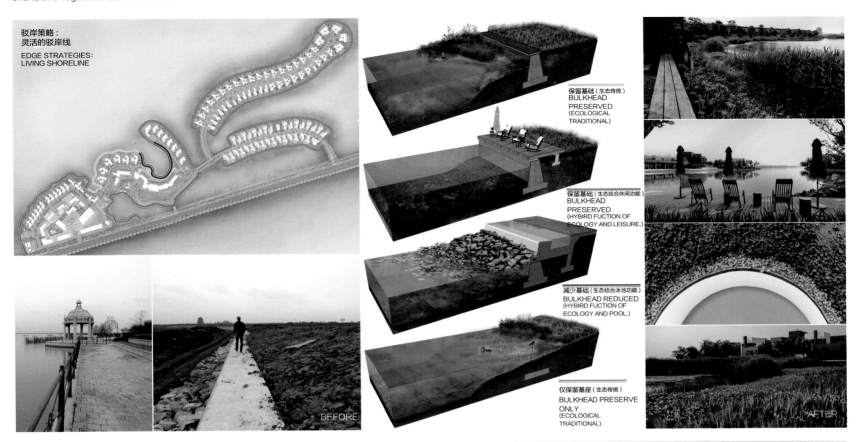

杭州萧山 前湾国际社区·无界公园
XIAOSHAN HANGZHOU BAY UNBOUNDED PARK

> 评委点评　JURY COMMENTS
>
> 本项目引入"无界"概念，将住宅社区融入公共空间，创造开放社区公园；以城市景观基础设施的姿态连接人与自然间的动态关系，成为居住景观设计标杆。
>
> With boundaryless design concept, this project integrates the residential community into the public space to create an open community park. On the other hand, it connects the dynamic relationship between human and nature through the urban landscape infrastructure, making it a benchmark of the residential landscape design.

第五届地产设计大奖·中国　景观设计 – 金奖
Landscape Design | Gold Award of the 5th CREDAWARD

业主单位：仁恒置地 & 香港置地	Owner: Yanlord Land & HongKong Land
规划设计：上海柏涛建筑设计有限公司	Planning Design: PTA
建筑设计：上海柏涛建筑设计有限公司	Architectural Design: PTA
景观设计：上海广亩景观设计有限公司	Landscape Design: Shanghai GM Landscape Design Co., Ltd.
室内设计：梁志天设计集团　大观建筑设计 DAGA Architects	Interior Design: Steve Leung Design Group　DAGA Architects
照明设计：南京弗思特照明设计有限公司	Lighting Design: FORSTER
幕墙设计：中国联合工程有限公司	Facade: China United Engineering Corporation Limited
项目地点：中国·浙江·杭州	Location: Hangzhou, Zhejiang, China
景观面积：27 961 m²	Landscape Area: 27,961 m²
建成时间：2018 年	Date of Completion: 2018
设计周期：9 个月	Cycle of Design: 9 months

当现有的社区都在打着"安全、私密、尊享"的口号时,前湾国际社区用敞开怀抱的"无界公园",向所有人发出信号——这里是我们共同的家,向世人传达开放、恒久、成长的无界生活理念。它是一个真正可以称得上"公园"的空间。

When the existing communities go publicity with the slogan of 'Safety, Privacy, Enjoyment', the Qianwan International Community Unbounded Park (QICUP) makes an announcement with the 'Boundless Park': This is our shared home. It aims to convey the concept of open, lasting, growing, and boundless life to the world. It deserves the name of 'park'.

城市界面 / Urban interface

社区样板 / Community model

自然公园 娱乐体验 / Nature park Enterainment experience

无界公园位于萧山区钱塘江畔，景观设计吸纳了以勇于闯荡、敢为人先为精神内核的"钱塘江水文化"，以"钱江潮"为出发点，提取文化元素，并以此构成统一设计语言注入景观布局及其细节当中。

The QICUP is located on the bank of Qiantang River in Xiaoshan District. Its landscape design absorbs the Water Culture of Qiantang River. The core spirit of that culture is brave and pioneering. The design takes the 'Qianjiang Tide' as the starting point to extract the cultural elements, forming a unified design language which are used in the landscape layout and its details.

04 室 内 设 计　INTERIOR DESIGN

01　香港希慎广场 —————————————————————— 254
　　　HYSAN PLACE, HONG KONG

02　格柏购物中心 —————————————————————— 260
　　　GERBER SHOPPING MALL

03　京都四季酒店全日餐厅 ————————————————— 266
　　　BRASSERIE RESTAURANT FOUR SEASONS KYOTO

274 ---------- 华侨城苏河湾上海宝格丽酒店 ● **04**
OCT SUHE CREEK BVLGARI HOTEL

284 ---------- 西打磨厂共享际 ● **05**
GRINDING FACTORY 5LMEET

香港希慎广场
HYSAN PLACE, HONG KONG

 评委点评   JURY COMMENTS

希慎广场可谓是"螺蛳壳里做道场"的典范，在单层面积非常有限的条件下把十几层的商业设计做活，并且还能设计出空中花园，实现了空间价值利用的最大化。

Hysan Place at Lee Gardens can be described as a model of 'doing dojo in a snail shell' which means doing a big thing in a very small place. For example, with a very limited area of a single story, the design of this project not only makes the businesses involved in more than a dozen floors blooming but also builds a sky garden, maximizing the use of space.

第一届地产设计大奖·中国　室内设计 – 金奖
Interior Design | Gold Award of the 1st *CRED*AWARD

业主单位：希慎兴业	Owner: Hysan Development Company Limited
室内设计：Benoy 贝诺	Architectural Design: Benoy
项目地点：中国·香港	Location: Hong Kong, China
商业面积：42 000 m²	Retail Space: 42,000 m²
建成时间：2013 年	Date of Completion: 2013

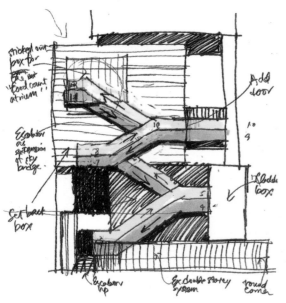
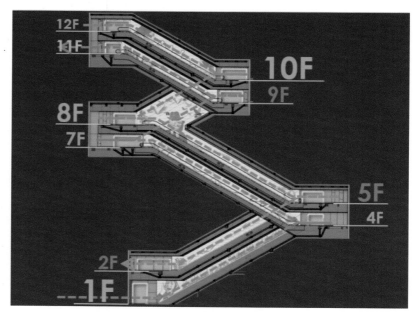

希慎广场位于繁忙的商业区铜锣湾，作为大中华区域首个获得LEED最高白金级别认证的购物中心发展项目，这座高17层的购物中心已成为该区域的新地标。其可持续环境设计包括：中央天窗，四楼的空中花园，精心挑选的材料，先进的通风系统、节水设备、低耗能装置，还将有两百年历史的柚木进行回收再利用。

Hysan Place is located in the busy business district of Causeway Bay. As the first shopping mall development project in the Greater China region to obtain the highest LEED platinum certification, this 17-storey shopping mall has become a new landmark in the area. Its sustainable environmental design includes: a central skylight, a sky garden on the fourth floor, and carefully selected materials, water-saving equipment, low energy consumption devices, and two-hundred-year-old teak wood for recycling.

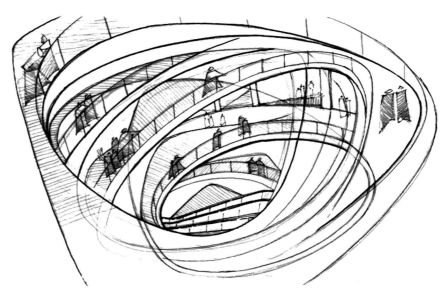

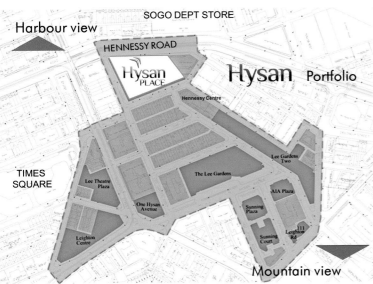

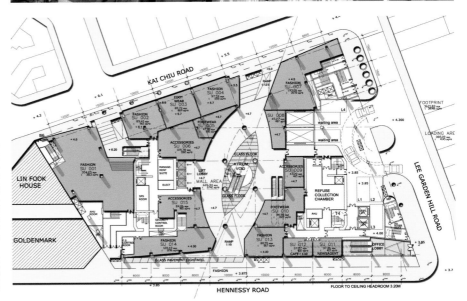

设计师精心挑选的材料营造出独特的美学效果,赋予希慎广场极具特色的绿色环保身份。节能照明、节水设备和最先进的新风系统塑造出大胆重叠的几何图案和素雅的单色调。

The carefully selected materials create a unique aesthetic effect, giving Hysan Place a distinctive green identity. Energy-saving lighting, water-saving equipment and the most advanced fresh air system create bold overlapping geometric patterns and elegant monotones.

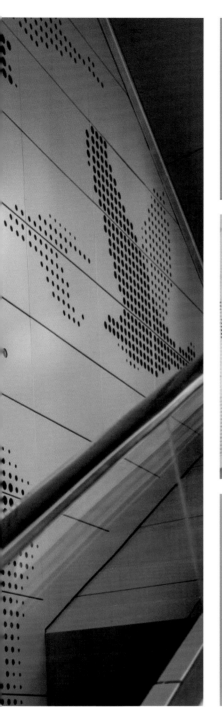
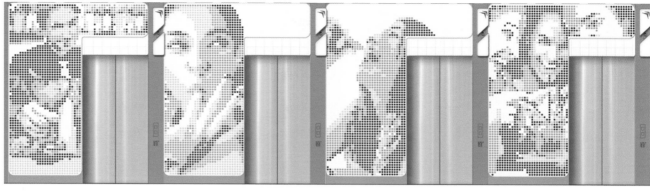

格柏购物中心
GERBER SHOPPING MALL

> **评委点评　JURY COMMENTS**
>
> 本项目的单色简洁设计凸显内饰的雅致，衬托商铺的缤纷色彩。清晰的导向系统和令人印象深刻的装饰元素避免给人奢华高傲的印象，以吸引项目定位的各层级客户群。
>
> The simple design with single color highlights the interior elegance to make its shops rich and colorful. Moreover, it's clear guiding system and impressive decoration avoid the luxurious and arrogant impression, meeting its demand to attract all kinds of customers.

第三届地产设计大奖·中国　室内设计 – 金奖
Interior Design | Gold Award of the 3rd CREDAWARD

业主单位：凤凰房地产开发有限公司	Owner: Phoenix Real Estate Development GmbH
室内设计：IFG 伊波莱茨建筑设计	Interior Design: Ippolito Fleitz Group-Identity Architects
项目地点：斯图加特	Location: Stuttgart
项目规模：120 000 m²	Project Scale: 120,000 m²
室内面积：5 500 m²	Interior Area: 5,500 m²
建成时间：2014 年	Date of Completion: 2014
设计周期：44 个月	Cycle of Design: 44 months

本项目将格伯中心打造成斯图加特的地标性建筑。公共区域的单色简洁设计凸显内饰的雅致，同时也避免了与商铺之间产生色彩冲突，争夺访客的眼球。

The GERBER Shopping Center is positioned to be a landmark building in Stuttgart. The design of the public areas in the monochromatic color and simple style highlights the elegance of the interior, while deliberately avoiding the visual competition with the facades of individual shop, and making the visitors focus on the shops.

· **融汇贯通，室内设计与导向系统完美结合。**
提取自格柏中心标志设计的椭圆形状和室内设计中的环形主题构成了整个导向系统设计的视觉主体，经变化调整被运用到购物区、地下车库和承租区的各种细节中，如立柱、吊式指示牌、导向系统的发光背景、地下车库的坡道和车位。导向系统中图形符号的设计语言符合时代美学，进一步强调出格柏中心的档次追求，及其独树一帜的个性特点。

· **Perfect rotations. What an interior design and orientation system can join hands.**
The entire orientation system for the mall, underground car park and tenant areas was designed and implemented using two graphic leitmotifs: an elongated hole from the mall's logo and the rings that are so much a part of the interior architecture. The graphic leitmotifs can be found in different variations on elements such as pillars, ceiling signage and the illuminated walls of the orientation system, as well as on the ramp and in the parking spaces of the underground car park.

京都四季酒店全日餐厅
BRASSERIE RESTAURANT FOUR SEASONS KYOTO

> 评委点评　JURY COMMENTS
>
> 这一餐厅通过克制的设计语言和手法展现了日本的文化。同一符号的连续使用使空间体验实现了立体的升华。室内外环境的延续则在精神层面给人带来历史的回忆。
>
> This restaurant presents Japanese culture through restrained design language and techniques. Besides, the repeated use of the same symbol creates 3D space experience. On the other hand, the continuation of the indoor and outdoor environment brings people historical memories.

第三届地产设计大奖·中国　室内设计 – 金奖
Interior Design | Gold Award of the 3rd *CREDAWARD*

业主单位：京都东山酒店资产 TMK	Owner: Kyoto Higashiyama Hospitality Assets TMK
室内设计：Kokaistudios	Interior Design: Kokaistudios
摄　　影：Seth Powers	Photographer: Seth Powers
项目地点：日本·京都	Location: Kyoto, Japan
项目规模：870 m²	Project Scale: 870 m²
室内面积：870 m²	Interior Area: 870 m²
建成时间：2016 年	Date of Completion: 2016
设计周期：24 个月	Cycle of Design: 24 months

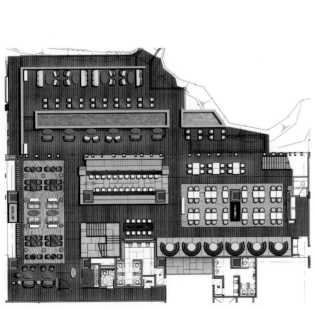

酒店依山傍水，紧临受联合国教科文组织保护的庙宇和历史悠久的积翠园。餐厅位于到达大堂和花园之间，是主要的公共空间，并因此成为整个酒店最具代表性的空间。

设计师把强烈的建筑手法带入室内设计中，并创新地运用材料和灯光营造人与空间柔和中有惊喜的微妙感觉。各类手工打造的细节以最精细的状态呈现出来与空间的昼夜光景产生交互作用，使空间整体呈现强大的表现力。

Surrounded by mountains and water, the hotel is close to the UNESCO-protected temple and the historic Ikeniwa Garden. The restaurant, which is the core public space as well as the most representative space in the entire hotel, is located between the arrival lobby and the garden.

The designer brings a strong architectural technique into the interior design, and innovatively uses materials and lights to create the unexpected subtle emphatic feeling between the space and the people. All kinds of hand-made details are presented in the finest state, interacting with the daylight and nightlight of the space, and revealing an incredibly expressive power.

为了呈现此独特项目，设计师以传统日式家具灵感加以追求舒适、现代的态度，定制了一系列格调优雅的家具。其中包括受当地传统竹编工艺启发而设计的蚁巢意象的标志性座椅，既符合空间的个性，又与户外联系了起来。

材料方面，石材表面为手工打磨，使其在视觉和触觉上均十分柔和；木饰表面以传统日式户外木的手法处理；特殊金属材料表面则由日本工匠制成，展现金属不同寻常的原始韵味。

In order to present this unique project, the designer customized a series of elegant furniture featured with the style of traditional Japanese furniture and the comfortable and modern traits, such as the iconic seats which were crafted as large wooden nests thanks to the inspiration brought by the local traditional bamboo weaving craftsmanship. Those seats not only match the space character, but also link to the outdoors.

When it comes to its materials, the stone surfaces were hand-hammered to make them visually soft and be felt softly; the wooden surfaces were treated in the same technic as the traditional Japanese outdoor timbers; the surfaces of the special metal materials were done in cooperation with Japanese craftsmen to reveal the unusual and original artistic effects.

华侨城苏河湾上海宝格丽酒店
OCT SUHE CREEK BVLGARI HOTEL

评委点评　JURY COMMENTS

苏河湾宝格丽酒店，由华侨城上海置地设计管理团队主持设计管理，室内设计融合了当代意大利的时尚理念和上海历史建筑保护的文化传承。历史风貌的专业化研究和修缮、历史文脉延续性的呈现、整体化的设计和当代化的再现，完美演绎了建筑风貌最辉煌的历史。

The Bvlgari Hotel at Suhewan is hosted by the Design Team in OCT (Shanghai)Real Estate Co., Ltd. The Interior design incorporates the Italian contemporary pioneering concepts with the technical experience of Shanghai historical building preservation. The professional repair of historical elements, the continuous presentation of historical context, the integrated design and the contemporary reappearance all perfectly interpret the most glorious history of the building itself.

第四届地产设计大奖 · 中国　室内设计 – 金奖
Interior Design | Gold Award of the 4th CREDAWARD

业主单位：华侨城（上海）置地有限公司	Owner: OCT Land (Shanghai) Investment Ltd.
规划设计：Foster+Partners	Planning Design: Foster+Partners
建筑设计：Foster+Partners	Architectural Design: Foster+Partners
傲地国际建筑设计有限公司	Lmk architects
华东都市建筑设计研究总院	ECADI
上海联创建筑设计有限公司	UDG
景观设计：Landscape Design Inc.	Landscape Design: Landscape Design Inc.
室内设计：Antonio Citterio Patricia Viel Interiors S.r.l	Interior Design: Antonio Citterio Patricia Viel Interiors S.r.l
BHD Consulting Limited.	BHD Consulting Limited.
照明设计：Project Lighting Design Pte Ltd.	Lighting Design: Project Lighting Design Pte Ltd.
项目地点：中国 · 上海	Location: Shanghai, China
项目规模：234 091 m²	Project Scale: 234,091 m²
建筑面积：234 000 m²	Floor Area: 234,000 m²
景观面积：37 000 m²	Landscape Area: 37,000 m²
室内面积：23 800 m²	Interior Area: 23,800 m²
建成时间：2018 年	Date of Completion: 2018
设计周期：60 个月	Cycle of Design: 60 months

华侨城苏河湾一街坊项目是上海城市中心新地标，世界级滨水城市复兴典范。其以现代的设计手法，将迎宾大道水纹铺装肌理延续到整个广场，再现当地特色的滨水文化；以现代的镜面水景作为光影载体，将历史时刻倒映在广场上。酒店花园采用镜面黑色石材景墙，形成另一个时间维度的倒影，让休憩于花园中的客人感受到历史底蕴的传承。

华侨城苏河湾东区项目占地面积41 984.5 m²，北至天潼路，东至河南北路，西至山西北路，南至北苏州路 南拥苏州河约200 m长河岸线，东临地铁10号线，北临地铁12号线，地块东北角有优秀历史保护建筑——原上海市总商会。该项目集商业区、酒店、住宅、办公区、历史建筑于一体，总建筑面积22.9万 m²。

OCT SUHE CREEK District project is a new landmark in the center of Shanghai city and a renewed model of the world-class waterfront city. With the modern techniques, the design of this project extends the texture of the water pavement in Yingbin avenue to the whole square, presenting the local waterfront culture. The design takes the modern mirror water scene as the light carrier to reflect the history on the square. The hotel garden uses the mirrored black stone landscape wall as the medium, forming another dimension of reflection, making the guests sitting in the garden feel the historical heritage.

With an area of 41984.5 m², the Overseas Chinese Town Suhewan East District is at the Tiantong Road in its north, at the north Henan Road in its east, at the North Shanxi Road in its west, at the North Suzhou Road in its south; it owns 200m river line along the Suzhou River in its south, is closed to the Metro 10# line in its east and the Metro 12# line in its north. The site is near the outstanding historical protected building-the former Shanghai General Chamber of Commerce in its northeast corner. The project sets commercial area, hotel, residence, office, historic buildings in one, with a total construction area of 0.229 million m².

总商会广场以百年总商会建筑主立面为原形,抽象演化;以不同颜色石材拼接及线条灯勾勒;以铺装倒影的方式展现建筑立面。地面铺装与草坪的融合渐变手法呈现历史的延续与发展。居住在酒店的客人将直观地感受到历史文化与现代生活的融合。

Taken the main facade of the century-old building, Shanghai Chamber of Commerce, as its original shape, the Plaza of Shanghai Chamber of Commerce is abstracted and stripped. With different colors of stone splicing and line lights outlining, the facade of the building is displayed on the pavement of the square. Besides, the junction of the pavement and the lawn is expressed in a gradually changing technique to present the continuation and development of history to let its guests living in the top floor intuitively feel the reflection of history melting the present.

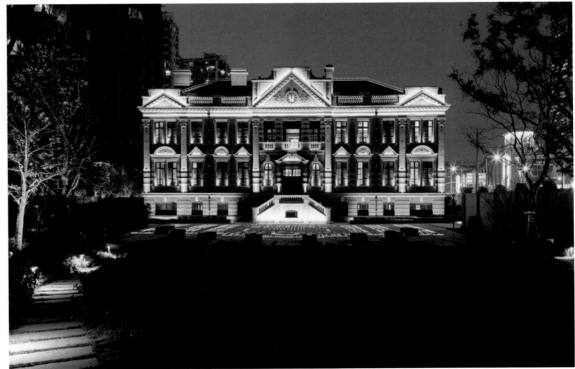

T1楼是40层、150米高的超高层商办综合楼。低区和中高区4~32层为公寓式办公区，高区34~39层为全球第四家顶尖奢侈品牌BVLGARI酒店的客房层，共82间客房。塔楼顶部会所配有意大利餐厅和酒吧，令客人尽享黄浦江景与陆家嘴风光。

With the height of 150m and 40 floors, the tower 1 (T1) is a commercial office building. The low area of this tower covering from the 4th to 32nd floor is the offices with the style of apartment, the high area of this tower covering from the 34th to 39th floor is the 82 guest rooms of BVLGARI Hotel, the 4th top luxury hotel brand in the world. The club in the top of this tower is equipped with an Italian restaurant and a bar, letting its guests enjoy the view of Huangpu River and Lujiazui landscape.

酒店客房设计匠心独运，轩敞疏朗，糅合了古典与当代，中国与意式文化。例如黑漆茶托搭配托斯卡纳鱼肚白大理石桌面，巧借东西方元素，共同打造摩登设计效果。酒店全景视野将外滩标志性景观、浦东摩登天际线一并收入。酒店客房家具及照明由Antonio Citterio Patricia Viel亲自挑选和定制设计，并由享誉全球的意大利品牌量身定制，以此确保风格的统一性。室内整体以哑光黑色花岗岩和浅色系石材为主，细部设计辅以精心打磨过的柚木和黄铜，二者相得益彰，用细节打造低调奢华感。酒店大面积选用黑色系的墙面和皮质沙发，营造出高品质的室内环境。

The design of the guest rooms is featured with ingenuity, openness, capaciousness, and the mixed-culture which refers to the combination of the classical culture with the contemporary one as well as the mixture of the Chinese culture and the Italian one. For example, the black lacquered saucer is paired with the Tuscan fish maw white marble tabletop, which skillfully absorbs the eastern and western elements to create a modern effect. The panoramic view of the hotel captures the iconic Bund landscape and the modern skyline of Pudong. The furniture and lights of hotel room were personally selected and designed by Antonio Citterio Patricia Viel, and customized by the world-renowned Italian brand to ensure the unity of its style. The interior is mainly made of matte black granite and light-colored stone, and the detailed design is accompanied by the carefully polished teak and brass to create a decent luxury image. The hotel is largely occupied with black walls and leather sofas, creating a high-quality indoor environment.

西打磨厂共享际
GRINDING FACTORY 5LMEET

" 评委点评　JURY COMMENTS

项目地处北京历史文化保护区域内，利用遗产建筑做底，在此上面做加法，设计手法克制且简练。项目力图在较小的空间实现功能最大化，虚实处理灵活，住户可感受到非常平和且清净的时空背景。

Located in the historical and cultural protection area in Beijing, the project is based on the heritage building to build up modestly, concisely. In addition, the project strives to maximize the space out of a small one with a delicate skill, making its residents feel peaceful and clean.

第五届地产设计大奖 · 中国　室内设计 – 金奖
Interior Design | Gold Award of the 5th *CRED*AWARD

业主单位：优享创智（北京）科技服务有限公司	Owner: 5Lmeet
室内设计：大观建筑设计 DAGA Architects	Interior Design: DAGA Architects
摄　　影：史云峰	Photographer: Yunfeng Shi
项目地点：中国 · 北京	Location: Beijing, China
建筑面积：680 m²	Floor Area: 680 m²
景观面积：160 m²	Landscape Area: 160 m²
室内面积：520 m²	Interior Area: 520 m²
建成时间：2017 年	Date of Completion: 2017
设计周期：1 个月	Cycle of Design: 1 month

1. 接待	1. Reception
2. 中央厨房	2. Central Kitchen
3. 洗衣房	3. Laundry
4. 设备间	4. Equipment Room
5. 锅炉房	5. Boiler Room
6. 餐厅	6. Restaurant
7. 卫生间	7. Bathroom
8. 酒吧	8. Bar
9. 厨房	9. Kitchen
10. 庭院	10. Courtyard
11. 客厅	11. Living Room
12. 客房卫生间	12. Guest Room Bathroom
13. 卧室	13. Bedroom

1. 接待 Reception
2. 洗衣房 laundry
3. 庭院 Courtyard
4. 客房 Guest Room

公寓由三个院子相连而成，从一面拱形木质大门走进院子，左侧为前台接待区，右侧为中央厨房。厨房对着庭院里的古树，为了将庭院里的古树完好地保留下来，平面布局巧妙地避开了古树的位置，用曲面玻璃幕墙把厨房与庭院分隔开。

The apartment is connected by three yards, with the entrance made of an arched wooden gate leading into the courtyard, which is divided into the front reception area on the left and the central kitchen on the right. The kitchen faces the ancient trees in the courtyard. These ancient trees are deliberately kept out of the plane layout in order to keep them intact in the courtyard, and the kitchen is separated from the courtyard with the curved glass curtain wall.

玻璃自身的透明特性不会遮挡原有建筑物,不仅符合人们对阳光的追求,更能够从视觉上以及空间原理上使得原有古建筑不会被削弱,所以整个院子的加建部分都是采用玻璃盒子的形式,再加上光线和人的运动,整个空间被赋予了活力。阳光穿过通透的玻璃倾泻到公寓内,给公寓内的一切带来温暖和灵动之感。

The transparency of the glass means that it will neither block the much--desired sunshine nor impair the original look of the building, creating a visually and spatially balance in relation to the old building. Therefore, the extended part of the whole yard is built in the form of a glass box. With light flittering through and light and people moving around in it, the entire space is full of vitality. When sunlight pours through the transparent glass into the apartment, it brings warmth and energy to everything in the apartment.

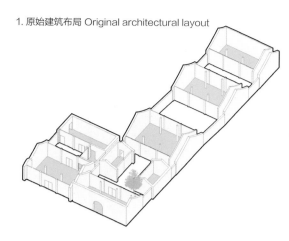

1. 原始建筑布局 Original architectural layout

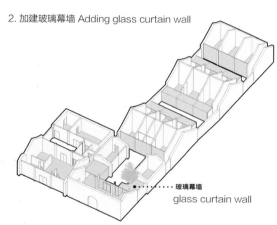

2. 加建玻璃幕墙 Adding glass curtain wall

玻璃幕墙 glass curtain wall

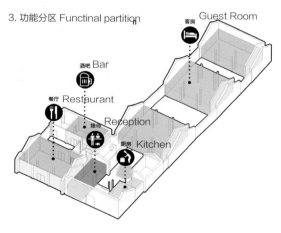

3. 功能分区 Functinal partition

客房 Guest Room
酒吧 Bar
餐厅 Restaurant
接待 Reception
厨房 Kitchen

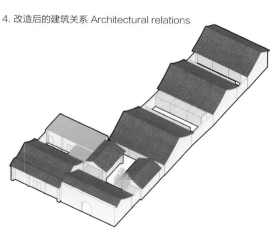

4. 改造后的建筑关系 Architectural relations

接待区、中央厨房、餐厅和酒吧都采用了玫瑰金色的金属板,这些公共空间保留了建筑原本的砖墙,使酷炫、光滑、金黄的金属板与沧桑、粗糙、灰暗的旧砖墙形成气质上的反差,新与旧相互产生碰撞,拉开时间上的层叠。

Rose-gold metal plates are used in reception, central kitchen, dining room and bar and the building's original brick walls are retained in these public spaces, making the cool, smooth, golden metal plates stand in stark contrast to the shabby, rough, gray old brick wall. This is indeed the collision between the new and old.

05 照 明 设 计 LIGHTING DESIGN

01 ● 长沙梅溪湖国际文化艺术中心 — 294
　　　MEIXI LAKE INTERNATIONAL CULTURE ART CENTER

长沙梅溪湖国际文化艺术中心
MEIXI LAKE INTERNATIONAL CULTURE ART CENTER

评委点评　JURY COMMENTS

设计师为一座高难度的建筑提供了完美的照明方案，使灯光与空间达到高度的统一，相得益彰。

The designer provides the perfect lighting solution for a difficult building, so that the lighting and the space achieve a high degree of unity and complement with each other.

第五届地产设计大奖·中国　照明设计 – 金奖
Lighting Design | Gold Award of the 5th *CREDAWARD*

业主单位：长沙梅溪湖实业有限公司	Owner: Changsha Meixihu Industry Co., Ltd.
建筑设计：扎哈·哈迪德建筑事务所	Architectural Design: Zaha Hadid Architects
照明设计：bpi	Lighting Design: bpi
项目地点：中国·湖南·长沙	Location: Changsha, Hunan, China
建筑面积：124 940 m²	Floor Area: 124,940 m²
室内面积：96 973 m²	Interior Area: 96,973 m²
建成时间：2018 年	Date of Completion: 2018
设计周期：48 个月	Cycle of Design: 48 months

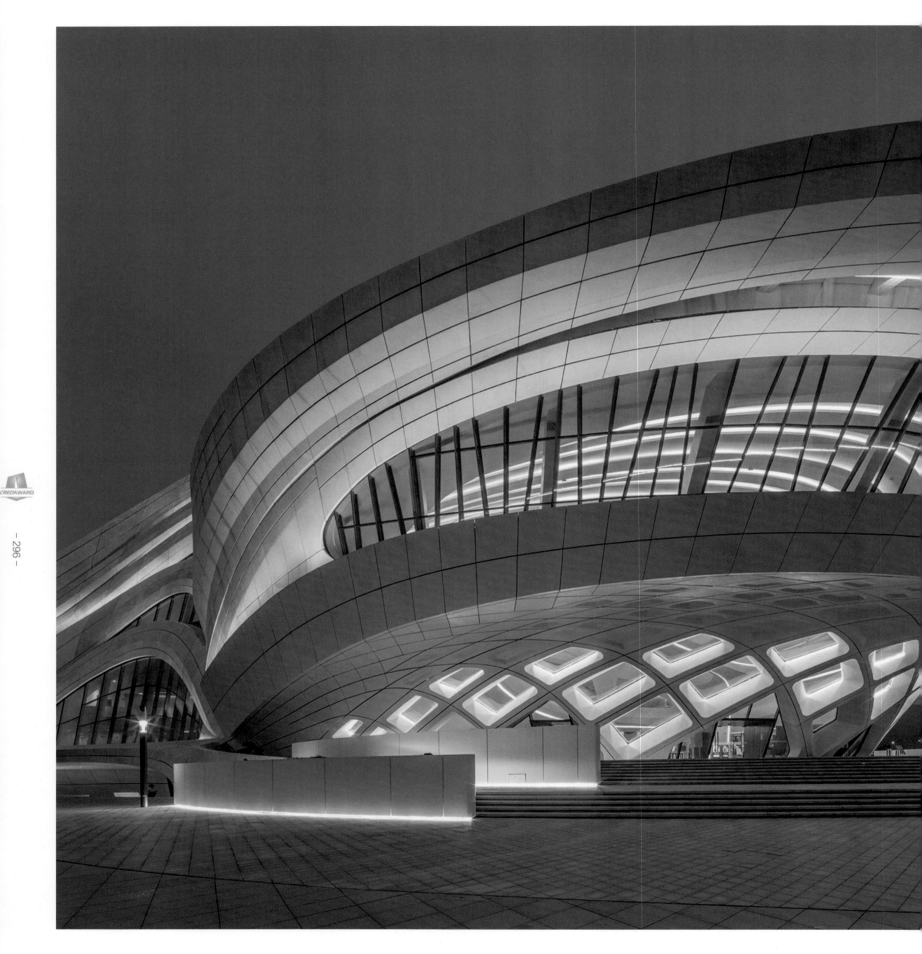

这是一座充满雕塑感的柔美建筑，扎哈的设计赋予了建筑空间永恒流动的生命。照明设计在项目中始终保持一种极致的纯粹。空间中的灯具不是独立地存在，而是与室内空间完美地融合为一体。对于形体复杂、流动扭转且功能多样、形态各异的建筑空间而言，这无疑是一项非常艰巨的挑战。

在这样一个永恒流动的空间里，为保证灯光完美的质感和纯粹性，灯具的隐藏成为了最重要且不容瑕疵的一件事。照明设计师尝试使用线性灯槽这种单一的照明系统去解决所有不同空间的问题。

在多层挑高的前厅空间，设计师通过灯槽数量、开口大小、造型设计以及槽内材质的多样化转变，使得天花灯光满足到地面的照度要求。

在一些贯穿立面及天花同时伴随三维扭曲的灯槽处，采用二次间接反射灯槽节点设计，真正确保观众在任何视点都看不到灯具的存在。

This building is gentle, graceful, and featured with the sculptural character. Zaha Hadid's design gives the construction space an eternal and flowing life. The lighting design has always maintained a kind of extreme purity in the project. The lamps are not isolated, but are perfectly integrated with the indoor space, which poses a great challenge for the architectural space featured with the complex shapes, flowing twists, diverse functions, and various forms.

It is vital for the lamps to be hidden to ensure the perfect texture and purity of the lighting in such an eternally flowing space. In order to reach that end, the lighting designers try to use a single lighting system such as linear light troughs to solve all the problems of different spaces.

In the multi-story high-rise lobby space, in order to make the ceiling lighting meet the requirements of the illuminance on the ground, the following measure is taken: change the number of lamp troughs, the size of the opening, the shape design and the materials in the trough.

The light troughs with 3D distortion, which penetrate the facade and the ceiling, are designed in the secondary indirect reflection light trough nodes to ensure the lamps are invisible.

照明设计希望运用线性灯光来引导并串联观众的情绪,使观众从室外进入前厅时被宏伟的空间感动,行进到环廊在柔和的过渡性灯光中等待。那一刻,观众感受到空间戏剧化以及震撼人心的氛围,对于空间的热爱转化成对于即将开演剧目的期待。

The aim of the lighting design is to use linear lighting to guide and have an emotional effect on the audience. In other words, when entering the lobby from the outside, the audience are moved by the magnificent space in the first place, then they move to the corridor to wait and be soaked in the soft transitional lighting. At that moment, the audience are deeply moved by the surroundings, looking forward to the upcoming program.

设计师将线性灯光作为设计语言铺展在流动的空间形态当中,不同细节变化的展现也演绎着不同的功能设置和空间布局。

为尽可能呈现完美的流星,照明设计需要一个既能精准控制单点跑动又能适应狭小三维曲面安装的,从而形成连续灯槽效果的灯具。在有限的技术限制下,设计师经过反复调试和改进配光,专门研发了一种特殊的串灯,并良好地应用于照明设计当中,为观众缔造绝佳的照明体验。

With the linear lighting as a design language spreading in the flowing space, the different details illustrate the different functional settings and spatial layouts.

In order to make the lights shine like shooting star does as much as possible, the luminaire is needed, which can precisely control the single-point movement and adapt to the installation of the narrow 3D curved surfaces, forming the effect of a continuous light trough. Due to the undesired technology, the designer has to have developed a special string of lights after repeatedly adjusting the light distribution and improving it, which works well during lighting, creating an excellent lighting experience for the audience.

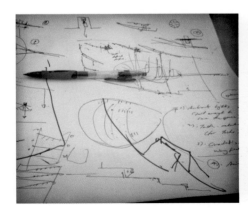

CREDAWARD 地产设计大奖·中国 金银奖年鉴（2014—2019）
GOLD&SILVER AWARDS OF CREDAWARD CATALOGUE(2014-2019)

01 公 建 项 目 PUBLIC CONSTRUCTION PROJECT

第一届金奖

1st GOLD AWARD

上海环贸广场 IAPM
IAPM SHANGHAI

广州 W 酒店及公寓
W GUANGZHOU HOTEL & RESIDENCES

天友绿色设计中心
TENIO GREEN DESIGN CENTER

北京百度科技园
BEIJING BAIDU TECHNOLOGY PARK

第二届金奖

2nd GOLD AWARD

杭州支付宝大厦
HANGZHOU ALIPAY BUILDING

佛山岭南天地商业街区
FOSHAN LINGNAN TIANDI

成都国际金融中心
CHENGDU INTERNATIONAL FINANCE SQUARE (IFS)

南昌绿地紫峰大厦
NANCHANG GREENLAND ZIFENG TOWER

虹桥天地
THE HUB

上海创智天地综合社区
SHANGHAI KNOWLEDGE & INNOVATION COMMUNITY

外滩 SOHO
BUND SOHO

● 第三届金奖

杭州西子湖四季酒店
FOUR SEASONS HOTEL HANGZHOU AT WESTLAKE

3rd
GOLD
AWARD

漕河泾现代服务业集聚区二期（三）
CAOHEJING THREE CUBES

西双版纳万达文华度假酒店
WANDA VISTA XISHUANGBANNA

北京绿地中心
BEIJING GREENLAND CENTER

天环广场
PARC CENTRAL

● 第四届金奖

明日希望
MEDIACITYUK TOMORROW

4th
GOLD
AWARD

响沙湾莲花酒店
XIANGSHAWAN DESERT LOTUS HOTEL

保利国际广场
POLY INTERNATIONAL PLAZA

丹寨万达小镇
DANZHAI WANDA TOWN

裸心 NAKED
NAKED

上海徐汇绿地缤纷城
SHANGHAI XUHUI GREENLAND BEING FUN CENTER

上海外滩金融中心
SHANGHAI BUND FINANCE CENTER

第五届金奖

5th GOLD AWARD

三亚海棠湾国际购物中心
SANYA HAITANG BAY INTERNATIONAL SHOPPING CENTER

平安金融中心
PING AN FINANCE CENTER

周大福金融中心
CTF FINANCE CENTER

船厂 1862
MIFA 1862

腾讯总部
TENCENT HEADQUARTERS

LCM 置汇旭辉广场
LCM PLAZA

万科时代中心 / 北京十里堡
VANKE TIMES CENTER / BEIJING SLP

武汉天地壹方北馆
NORTH HALL OF WUHAN TIANDI HORIZON

第四届银奖

4th SILVER AWARD

水晶大楼—明斯特保险公司总部
CRYSTAL—LVM5 HEADQUARTERS, MÜNSTER

银川韩美林艺术馆
YINCHUAN HAN MEILIN ART MUSEUM

郑州普罗理想国艺术文化中心
ZHENGZHOU PULUO IDEAL LAND—ART & CULTURE CENTER

世贸一期
THE NEW BUND WORLD TRADE CENTER (PHASE I)

枣庄市文体中心体育场
ZAOZHUANG SPORT AND CULTURE CENTER STADIUM

虹桥世界中心
HONGQIAO WORLD CENTER

第五届银奖

5th
SILVER
AWARD

曼海姆商学院研究与会议中心
MBS STUDY AND CONFERENCE CENTRE

长春水文化生态园
CHANGCHUN CULTURE OF WATER ECOLOGY PARK

荃湾体育馆
TSUEN WAN SPORTS CENTRE

盛港综合及社区医院
SENGKANG GENERAL AND COMMUNITY HOSPITAL

亚马逊星球
AMAZON SPHERES

第一届·金奖
1st GOLD AWARD

上海环贸广场 IAPM
IAPM SHANGHAI

业主单位：新鸿基地产
建筑设计：Benoy 贝诺
项目规模：47 000 m²
建筑面积：216 408 m²
建成时间：2012 年

Owner: Sun Hung Kai Properties
Architectural Design: Benoy
Project Scale: 47,000 m²
Floor Area: 216,408 m²
Date of Completion: 2012

Mixed-use ·综合体·

IAPM 的深夜购物体验拓展全新的商业模式，将林荫大道购物体验延伸至地下，打造成一系列模仿林荫大道风格的餐饮区，让这些餐饮"盒子"镶嵌在建筑体上，营造出别出心裁的空间形态。

The mall has been conceived as a day to night experience, am to pm. It expands a new business model, extending the boulvard shopping experience to the underground, creating a series of dining areas imitating the boulevard style, allowing these "boxes" to be embedded in the building to create a unique space.

第一届·金奖
1st GOLD AWARD

广州 W 酒店及公寓
W GUANGZHOU HOTEL & RESIDENCES

业主单位：合景泰富地产控股有限公司
建筑设计：许李严建筑师事务有限公司
建筑面积：106 504 m²
建成时间：2013 年
设计周期：84 个月

Owner: KWG Property Holding Ltd.
Architectural Design: Rocco Design Architects Ltd.
Floor Area: 106,504 m²
Date of Completion: 2013
Cycle of Design: 84 months

Hotel ·酒店·

该项目是一座包含精品酒店及服务式公寓的综合体，演绎紧凑式都市酒店的独特性。整体建筑加强和延续了街道的建筑面，内庭连接大街，开启了一扇引入空气、光线和城市活力的"城市窗户"。

W Guangzhou is a mixed-use consisting of a city hotel and serviced apartments, deducing the uniqueness of a compact urban hotel. The overall building strengthens the architectural surface of the street. The inner courtyard connects to the street, opening an 'urban window' that introduces air, light, and urban vitality.

天友绿色设计中心
TENIO GREEN DESIGN CENTER

业主单位：天友设计集团
建筑设计：天友设计集团
项目规模：5 700 m²
建筑面积：5 700 m²
建成时间：2013 年
设计周期：6 个月

Owner: Tenio Design Group
Architectural Design: Tenio Design Group
Project Scale: 5,700 m²
Floor Area: 5,700 m²
Date of Completion: 2013
Cycle of Design: 6 months

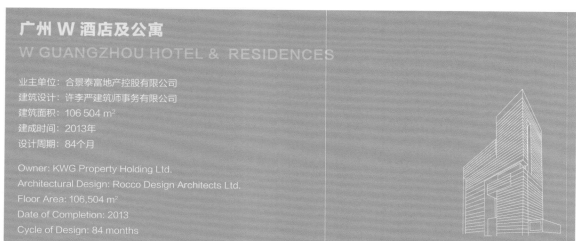

第一届·金奖
1st GOLD AWARD

Headquarter ·企业总部·

天友绿色设计中心以"问题导向的技术集成"为设计原则，从"被动优先，主动优化"出发，应用多功能的绿色技术，为多种实验性技术提供了应用平台和可靠数据，在运营中实现了优异的超低能耗性能。

Based on the principle of 'Problem-oriented Technology Integration', the design of Tenio Green Design Center, starting from the concept of "Passive Priority, Active Optimization", has achieved the ultra-low energy consumption performance with multi-functional green technology, providing an applied platform and reliable data for a variety of trial technologies, and presenting the excellent ultra-low energy consumption performance in operation.

第一届·金奖
1st GOLD AWARD

Industrial Plant ·产业园·

北京百度科技园
BEIJING BAIDU TECHNOLOGY PARK

业主单位：百度在线网络技术（北京）有限公司
建筑设计：维思平建筑设计
项目规模：75 839 m²
建筑面积：301 744 m²
建成时间：2014年（一期），2016年（二期）
设计周期：3个月

Owner: BAIDU INC.
Architectural Design: WSP ARCHITECTS
Project Scale: 75,839 m²
Floor Area: 301,744 m²
Date of Completion: 2014(phase I); 2016(phase II)
Cycle of Design: 3 months

本项目整体设计贯彻"简单可依赖"的百度理念，定位"绿色家园"。"浮岛"的概念充分考虑节能减排、低碳环保的绿色应用，实现高效且低成本运营。节能设计每年可节省75万度电，减少碳排放量66 645吨。

Baidu Science and Technology Park is designed to implement the Baidu concept of "Simple and Dependable" and aims to create a "green home". The concept of "Floating Island" considers the application of energy-saving and emission-reduction, eco-logical and low-carbon technologies to achieve high efficiency and low operating costs. It can save 750,000 kilowatt-hours of electricity for office buildings and reduce 66,645 tons of carbon emissions.

杭州支付宝大厦
HANGZHOU ALIPAY BUILDING

业主单位：杭州邮政科技实业有限公司
建筑设计：维思平建筑设计
项目规模：19 900 m²
建筑面积：95 712 m²
建成时间：2012年
设计周期：3个月

Owner:
Hangzhou Post Technology Industrial Co., Ltd.
Architectural Design: WSP ARCHITECTS
Project Scale: 19,900 m²
Floor Area: 95,712 m²
Date of Completion: 2012
Cycle of Design: 3 months

第二届·金奖
2nd GOLD AWARD

Headquarter ·企业总部·

支付宝大厦利用基地的自然条件和周边环境特点，沿街设置板式高层办公楼和公寓。L形的大楼将基地自然分成临街商业街区和相对安静的内庭，简洁如玉的形体使大楼成为街景的标志。

Based on the natural conditions of the site and the traits of its surroundings, Alipay Building lays out its slab-style high-rise offices and apartments along the street. The L-shaped building naturally divides the site into a commercial block facing the street and a relatively quiet inner courtyard, making the building a landmark in the street thanks to its simple but elegant shape.

第二届·金奖
2nd GOLD AWARD

Urban Regeneration ·城市再生·

佛山岭南天地商业街区
FOSHAN LINGNAN TIANDI

业主单位：瑞安房地产
建筑设计：Ben Wood Studio Shanghai 巴马丹拿集团
项目规模：55 020 m²
建筑面积：61 219 m²
建成时间：2013年
设计周期：36个月

Owner: Shui On Land
Architectural Design: Ben Wood Studio Shanghai
P&T Design and Engineering Limited.
Project Scale: 55,020 m²
Floor Area: 61,219 m²
Date of Completion: 2013
Cycle of Design: 36 months

岭南天地商业街区为满足采光和通风要求对建筑区域进行改建，以高品质的商业结构和丰富的文化元素实现本土与国际、传统与现代的完美融合，为佛山精心打造了一个优质的市中心综合发展项目。

Lingnan Tiandi is renovated to meet the requirements of daylighting and ventilation. The high-quality commercial structure and rich cultural elements realize the perfect integration of the local and the international, the tradition and the modernity, creating a high-quality down-town for Foshan.

成都国际金融中心
CHENGDU INTERNATIONAL FINANCE SQUARE (IFS)

业主单位：九龙仓集团有限公司
建筑设计：Benoy 贝诺；KPF建筑设计事务所
项目规模：210 000 m²（商业面积）
建筑面积：436 309 m²
建成时间：2014年
设计周期：48个月
Owner: Wharf Holdings Limited.
Architectural Design: Benoy
Kohn Pedersen Fox Associates(KPF)
Project Scale: 210,000 m² (Commercial Area)
Floor Area: 436,309 m²
Date of Completion: 2014
Cycle of Design: 48 months

Mixed-use · 综合体 ·

项目位于成都最繁华的商圈，"城中城"零售商业裙楼的建筑外观融合为一街景之一。国际甲级办公楼、五星级酒店、豪华住宅和商业广场一并组成高端的综合体目的地，四周被户外餐饮以及艺术展览馆所环绕。

Occupying a coveted location within the CBD, Chengdu IFS has been conceived as "a City within a City". The retail podium establishes a strong architectural presence for the development at street level. International Class A office buildings, five-star hotels, luxury residences and commercial plazas form a high-end complex destination, surrounded by outdoor dining and art exhibition halls.

南昌绿地紫峰大厦
NANCHANG GREENLAND ZIFENG TOWER

业主单位：绿地控股集团
建筑设计：Skidmore, Owings & Merrill (SOM)
建筑面积：209 058 m²
建成时间：2014年

Owner: GREENLAND HOLDING GROUP
Architectural Design: Skidmore, Owings & Merrill (SOM)
Floor Area: 209,058 m²
Date of Completion: 2014

Mixed-use · 综合体 ·

项目坐落在蓬勃发展的高新商务区内，与金融中心隔赣江相望。建筑中强大的斜肋构架遮阳系统和独特的"城市之窗"所界定的矩形体量，使这栋建筑成为南昌城市天际线中的一道独特风景线。

The project is located in the high-tech business district, facing the financial center across the Ganjiang River. The strong diagonal rib frame shading system and the rectangular volume defined by the unique "city window" in the building make this building to a unique landscape in the city skyline of Nanchang.

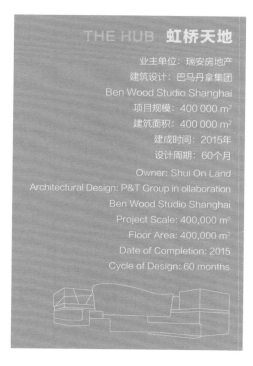

THE HUB 虹桥天地

业主单位：瑞安房地产
建筑设计：巴马丹拿集团
Ben Wood Studio Shanghai
项目规模：400 000 m²
建筑面积：400 000 m²
建成时间：2015年
设计周期：60个月
Owner: Shui On Land
Architectural Design: P&T Group in ollaboration
Ben Wood Studio Shanghai
Project Scale: 400,000 m²
Floor Area: 400,000 m²
Date of Completion: 2015
Cycle of Design: 60 months

· TOD ·

虹桥天地是集展示办公、购物、餐饮、休闲、娱乐、演艺为一体的一站式新生活中心，为虹桥商务区的7 500万人口提供商务、休闲、娱乐的新型地标平台，打造一个面向多元社群的活力社区。

The HUB is a one-stop new life center integrating office, shopping, leisure, entertainment, and performing arts. It provides a new landmark platform for business, leisure, and entertainment for the 75 million people in Hongqiao Business District and creates a diversified community.

第二届·金奖
2nd GOLD AWARD

Mixed-use Community ·复合功能社区·

创智天地充分响应上海"科教兴市"战略，以硅谷创新精神与左岸创意文化为灵感，打造城市转型发展中的"知识型社区"，使其成为杨浦区从"传统工业"向"知识创新"转变的标志性转折。

Responding to Shanghai "Revitalizing the City with Science and Technology" strategy, also inspired by Silicon Valley's innovative spirit and Left Bank's creative culture, KIC successfully developed a knowledge community. It marked the turning point for Yangpu District from Traditional Industrial Area to Knowledge Innovation Area.

上海创智天地综合社区
SHANGHAI KNOWLEDGE & INNOVATION COMMUNITY

业主单位：瑞安房地产
建筑设计：Skidmore, Owings & Merrill (SOM); Palmer & Turner; Atkins; Charpentier Architecture Design; Terry Farrell & Partners; Tongji Design Institute; BWSS; Gensler; TIANHUA; Leigh & Orange; Hassell; 3XN; AECOM; EMBT
规划设计：Skidmore, Owings & Merrill (SOM)
项目规模：490 000 m²
建筑面积：762 217 m²
建成时间：2015年
设计周期：24个月

Owner: Shui On Land
Architectural Design: Skidmore, Owings & Merrill(SOM); Palmer & Turner; Atkins; Charpentier Architecture Design; Terry Farrell & Partners; Tongji Design Institute; BWSS; Gensler; TIANHUA; Leigh & Orange; Hassell; 3XN; AECOM; EMBT
Planning Design: Skidmore, Owings & Merrill(SOM)
Project Scale: 490,000 m²
Floor Area: 762,217 m²
Date of Completion: 2015
Cycle of Design: 24 months

第二届·金奖
2nd GOLD AWARD

外滩 SOHO
BUND SOHO

业主单位：SOHO 中国
建筑设计：gmp·冯·格康，玛格及合伙人建筑师事务所
联合设计：华东建筑设计研究院总院
建筑面积：189 509 m²
建成时间：2015年
设计周期：60个月

Owner: SOHO China
Architectural Design: gmp·von Gerkan, Marg and Partners Architects
Joint Design: ECADI
Floor Area: 189,509 m²
Date of Completion: 2015
Cycle of Design: 60 months

Commercial Project ·商办项目·

外滩 SOHO 商业办公综合体以优雅的剪影收束了拥有上海"东方巴黎"盛誉的外滩天际线，延续万国建筑群的历史风格，为守护传承上海城市的地域文脉做出了贡献。

With the elegant silhouette, the SOHO commercial office complex on the Bund is the end of the Bund skyline that has given Shanghai the reputation of "Oriental Paris". It carries the historical style of the World Architecture, making contribution to protecting and inheriting the regional culture of the Shanghai.

第三届·金奖
3rd GOLD AWARD

Resort Hotel ·度假酒店·

项目将发掘景观因素和延续自然意境作为酒店的营造重点，精心思考和设计建筑群整体关系和庭院空间布局。围绕景观主轴两翼实现客房、庭院和宴会厅的铺展，叠山引水，如诗如画。

Focusing on the landscape and nature, the project carefully thinks about the relationship between the building group and the spatial layout of the courtyard. The guest rooms, courtyards and banquet halls are laid along the two wings of the main axis of the landscape, making this hotel picturesque.

杭州西子湖四季酒店
FOUR SEASONS HOTEL HANGZHOU AT WESTLAKE

业主单位：杭州金沙港旅游文化村有限公司
建筑设计：GOA 大象设计
建筑面积：43 537 m²
建成时间：2010年
设计周期：72个月

Owner: Hangzhou Jinsha Harbour Tourism and Cultural Village Co., Ltd.
Architectural Design: GOA
Floor Area: 43,537 m²
Date of Completion: 2010
Cycle of Design: 72 months

Office Plant · 办公园区 ·

设计理念在于建筑综合体的优雅感和同质性,单体组合将建筑群从异质化的周边环境中凸显出来。自然采光的地下空间实现了空间的流动贯通,与冷峻、凝练、优雅的建筑外形构成鲜明对比。

The design places emphasis on the character and elegance of the ensemble. The interaction between harmony and contrast is replicated in the design of the exterior surfaces. An undulating grass landscape provides access to the naturally lit basement floors. The landscape stands in flowing contrast to the static cubes in a harmoniously unifying manner, while at the same time emphasizing the cool austere elegance of the building volumes.

漕河泾现代服务业集聚区二期(三)
CAOHEJING THREE CUBES

业主单位:上海漕河泾开发区高科技园发展有限公司
建筑设计:gmp·冯·格康,玛格及合伙人建筑师事务所
合作设计:上海核工程研究设计院
项目规模:23 280 m²
建筑面积:90 650 m²
建成时间:2015年
设计周期:41个月

Owner: Shanghai Caohejing Hi-Tech Park Development Corporation
Architectural Design: gmp · von Gerkan, Marg and Partners Architects
Partner office: Shanghai Nuclear Engeneering Research & Design Institute
Project Scale: 23,280 m²
Floor Area: 90,650 m²
Date of Completion: 2015
Cycle of Design: 41 months

西双版纳万达文华度假酒店
WANDA VISTA XISHUANGBANNA

业主单位:大连万达集团股份有限公司
建筑设计:OAD 欧安地建筑设计公司
项目规模:150 000 m²
建筑面积:46 149 m²
建成时间:2015年
设计周期:29个月

Owner: DaLian Wanda Group
Architectural Design: o.ffice for a.rchitecture+d.esign(OAD)
Project Scale: 150,000 m²
Floor Area: 46,149 m²
Date of Completion: 2015
Cycle of Design: 29 months

Resort Hotel · 度假酒店 ·

项目位于云南省最南端的群山密林处,充分发掘当地热带环境资源和傣族文化特色,让建筑在内外空间上的呼应与交流延伸至精神层面,为游客提供一个隐于自然风景、私密幽静的度假乐土。

Wanda Vista Xishuangbanna is located in Xishuangbanna, one of the most emblematic regions of China, characterized for is natural wealth and cultural diversity. The architects take the traditional Dai architecture and planning as the inspiration to build a contemporary construction building and bridge it with its origins and surroundings.

北京绿地中心
BEIJING GREENLAND CENTER

业主单位:绿地控股集团
建筑设计:Skidmore, Owings & Merrill (SOM)
建筑高度:260 m
建筑面积:172 734 m²
建成时间:2016年
设计周期:16个月

Owner: GREENLAND HOLDING GROUP
Architectural Design: Skidmore, Owings & Merrill(SOM)
Architectural Height: 260 m
Floor Area: 172,734 m²
Date of Completion: 2016
Cycle of Design: 16 months

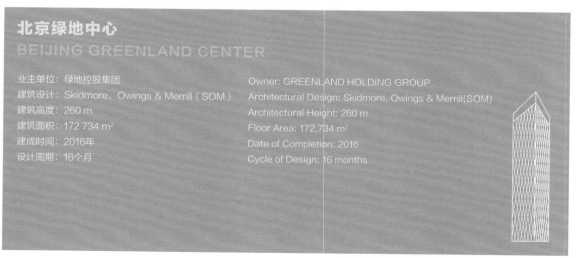

Commercial Project · 商办项目 ·

设计师秉持可持续发展和节能高效的设计理念,使北京绿地中心成为兼具视觉吸引力、高度灵活性和可持续发展性的综合属性城市项目。

Beijing Greenland Center focuses on sustainable development and energy-efficient design concepts, making it a comprehensive urban project with visual appeal, high flexibility and sustainable development.

3rd GOLD AWARD

· TOD ·

本项目不单是零售商场，更是绿色社区资产，多层次绿化为城市商业发展带来一座绿色公园。设计团队将大部分发展规划于地下，以此创造出一个广阔的地面绿化广场和全年无休服务社区。

The project is not only a retail mall but also a green community asset. The multi-level greening brings a green park for urban commercial development. The design team places most of the planning development in the underground to create a vast ground green plaza and serve community all year around.

天环广场
PARC CENTRAL

业主单位：新鸿基地产&广州新中轴建设有限公司
建筑设计：Benoy贝诺
项目规模：110 000 m²
建筑面积：110 000 m²
建成时间：2016年

Owner: Sun Hung Kai Properties & Guangzhou New Central Axis Construction Co., Ltd.
Architectural Design: Benoy
Project Scale: 110,000 m²
Floor Area: 110,000 m²
Date of Completion: 2016

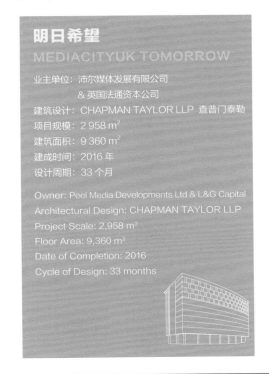

明日希望
MEDIACITYUK TOMORROW

业主单位：沛尔媒体发展有限公司
　　　　　&英国法通资本公司
建筑设计：CHAPMAN TAYLOR LLP 查普门泰勒
项目规模：2 958 m²
建筑面积：9 360 m²
建成时间：2016年
设计周期：33个月

Owner: Peel Media Developments Ltd & L&G Capital
Architectural Design: CHAPMAN TAYLOR LLP
Project Scale: 2,958 m²
Floor Area: 9,360 m²
Date of Completion: 2016
Cycle of Design: 33 months

4th GOLD AWARD

Mixed-use · 综合体 ·

明日希望是商业办公与酒店两种业态混合而成的综合体，属于英国媒体城一期开发范围。该楼达到英国建筑节能标准绿色建筑评估体系认证（BREEAM）优秀等级，是低碳可持续技术的典范之作。

The building in the name of the MediaCityUK Tomorrow, serving as the business office and hotel, belongs to the development scope of MediaCityUK Phase 1. The building has achieved the excellent level of the British Building Energy Efficiency Standard Green Building Assessment System Certification (BREEAM), enjoying the reputation of a model of low-carbon sustainable technology.

4th GOLD AWARD

Resort Hotel · 度假酒店 ·

项目位于鄂尔多斯库布其沙漠，PLAT ASIA探索出一种适用于沙漠生态的新型基础系统：采用钢板将流动性强的沙丘固定做基础，内外墙体用轻钢龙骨体系，轻薄的膜结构则作为独立的遮阳体系。

The hotel is located in Kubuqi Desert in Ordos. PLAT ASIA has invented a nouveau foundation system, making use of layers of steel panels to have the sands fixed in place, which is a custom to the special geological conditions in deserts. Moreover, instead of a sheer wall system, a precast steel skeleton system has been used to reduce the pressure on the foundation. This also allows the membrane system to be separately installed as a sun-shading device.

响沙湾莲花酒店
XIANGSHAWAN DESERT LOTUS HOTEL

业主单位：内蒙古响沙湾旅游有限公司
建筑设计：PLAT ASIA；UAA
项目规模：300 000 m²
建筑面积：48 000 m²
建成时间：2016年
设计周期：86个月

Owner: Inner Mongolia Xiangshawan Travel Co., Ltd.
Architectural Design: PLAT ASIA; UAA
Project Scale: 300,000 m²
Floor Area: 48,000 m²
Date of Completion: 2016
Cycle of Design: 86 months

保利国际广场
POLY INTERNATIONAL PLAZA

第四届·金奖
4th GOLD AWARD

业主单位：保利（北京）房地产开发有限公司	Owner: China Poly Real Estate Company Limited.
建筑设计：Skidmore, Owings & Merrill (SOM)	Architecture Design: Skidmore, Owings & Merrill (SOM)
结构设计：Skidmore, Owings & Merrill (SOM)	Structure Design: Skidmore, Owings & Merrill (SOM)
土木工程：Skidmore, Owings & Merrill (SOM)	Civil Engineering Design: Skidmore, Owings & Merrill (SOM)
高层建筑设计：Skidmore, Owings & Merrill (SOM)	High-rise Architecture Design: Skidmore, Owings & Merrill (SOM)
项目规模：23 612 m²	Consultant: 23,612 m²
建成时间：2017年	Date of Completion: 2017
设计周期：12个月	Cycle of Design: 12 months

Office Building · 办公项目 ·

建筑外围的折叠造型赋予其中国纸灯笼的特征，遍布项目场地的绿化景观和环保形态也让其成为新的办公塔楼典范。

The folding shape of the building's periphery gives it the characteristics of Chinese paper lanterns, and the green landscape and environmental protection forms covering all the site also make it a model for new office towers.

第四届·金奖
4th GOLD AWARD

丹寨万达小镇
DANZHAI WANDA TOWN

业主单位：大连万达集团股份有限公司
建筑设计：万达商业规划研究院有限公司
　　　　　上海力夫建筑设计有限公司
　　　　　重庆市设计院有限公司
项目规模：124 974 m²
建筑面积：50 000 m²
建成时间：2017年
设计周期：12个月

Owner: Dalian Wanda Group
Architectural Design:
Wanda Business Planning Research Institute Co., Ltd.
Shanghai LIFE Architectural Design Co., Ltd.
Chongqing Architectural Design Institute Co., Ltd.
Project Scale: 124,974 m²
Floor Area: 50,000 m²
Date of Completion: 2017
Cycle of Design: 12 months

Characteristic Town · 特色小镇 ·

项目位于贵州黔东南自治州丹寨县东湖西岸，以苗族特色建筑为基础，传承和发展丹寨历史文化与民族风情，是一座集"吃、住、行、游、购、娱、教"为一体的文化、非遗、养生旅游小镇。

The project is located on the west bank of East Lake in Danzhai County, Qiandongnan Autonomous Prefecture, Guizhou Province. It is based on the characteristic buildings of the Miao nationality to inherit and develop Danzhai's historical culture and ethnic customs. It is a tourist town that integrates "food, housing, travel, shopping, entertainment, and education", including culture, intangible cultural heritage and health preservation.

裸心 NAKED NAKED

业主单位：裸心 Naked
建筑设计：裸心 Naked；TIANHUA 天华；
项目规模：15 000 m²
建筑面积：15 000 m²
建成时间：2017年
设计周期：24个月

Owner: Naked
Architectural Design: Naked; TIANHUA;
Project Scale: 15,000 m²
Floor Area: 15,000 m²
Date of Completion: 2017
Cycle of Design: 24 months

第四届·金奖
4th GOLD AWARD

Resort Hotel · 度假酒店 ·

裸心堡位于浙江省德清莫干山，是基于城堡遗迹的复原改造项目。在尊重环境和历史风貌的基础上，项目强调可持续发展理念，引入西式度假设计思维，兼顾经济发展和区域品位，打造环境友好型顶级度假村。

Naked Castle, located in the Deqing Mogan Mountain, Zhejiang Province, is a rehabilitation project based on coastal ruins. Based on maintaining the deep historical and cultural background of the Mogan Mountain, the project focuses on the sustainable development concept. In the process of the design, the western style holiday design concept is introduced, the economy and regional quality are taken into consideration in order to create an environment-friendly top-class resort.

· TOD ·

本项目位于上海徐汇滨江地区，是以地铁 7 号线、12 号线换乘站为中心的 30 万 m² 的 TOD 综合开发设施，与地铁无缝对接，是上海与地铁站点接口最多的项目。

Located by the Huangpu River in Xuhui District, Shanghai, the project, owning 300,000 m² TOD comprehensive development facilities, is centered on the intersection of the subway Line 7 and Line 12, making it seamlessly to connect to the subway and have the most interfaces to the subway in Shanghai.

上海徐汇绿地缤纷城
SHANGHAI XUHUI GREENLAND BEING FUN CENTER

业主单位：绿地控股集团&上海绿地恒滨置业有限公司
建筑设计：株式会社日建设计
联合设计：华东建筑设计研究总院
项目规模：304 910 m²
建筑面积：21 925 m²
建成时间：2017年
设计周期：18个月

Owner: GREENLAND HOLDING GROUP &Shanghai Greenland Hengbin Properties Limited.
Architectural Design: NIKKEN SEKKEI LTD.
Joint Design: ECADI
Project Scale: 304,910 m²
Floor Area: 21,925 m²
Date of Completion: 2017
Cycle of Design: 18 months

上海外滩金融中心
SHANGHAI BUND FINANCE CENTER

业主单位：上海证大外滩国际金融服务中心置业有限公司   Owner: Shanghai Zendai Bund Int'l Finance Center Real Estate Co., Ltd.
建筑设计：Foster + Partners，Heatherwick Studio   Architectural Design: Foster + Partners, Heatherwick Studio
项目规模：426 073 m²   Project Scale: 426,073 m²
建筑面积：426 073 m²   Floor Area: 426,073 m²
建成时间：2017年   Date of Completion: 2017
设计周期：24个月   Cycle of Design: 24 months

· Mixed-use · 综合体 ·

外滩金融中心位于中国上海黄浦区，是外滩金融集聚带核心位置首个体验式复合型金融中心，汇集金融、商业、旅游、文化、艺术等多种功能，涵盖甲 A 级办公楼、购物中心、复星艺术中心、精品酒店四大业态。

located at Huangpu District, Shanghai, China, the Bund Finance Centre (BFC), as the first experiential mixed-use financial center at the core financial zone of the Bund, serves as multi-functions, such as finance, commerce, tourism, culture, and arts, covering Grade-A office buildings, shopping centers, Fosun Art Center, and a boutique hotel.

· Shopping Mall · 购物中心 ·

本项目位于三亚市海棠湾黄金地段，与亚龙湾和陵水湾相连，是占地约 19.2 万 m² 的一线海景用地。地上建筑主要功能为商场，地下建筑主要功能为办公、储藏、物流、停车，员工服务等。

The site of the project is in Haitang Bay, Sanya City, connected with Yalong Bay and Lingshui Bay. An area of about 192,000 m² of land for the first line of seascape. The main function above ground is shopping mall. The main functions underground are office, storage, logistics, parking, and staff services.

三亚海棠湾国际购物中心
SANYA HAITANG BAY INTERNATIONAL SHOPPING CENTER

业主单位：国旅（三亚）投资发展有限公司
建筑设计：法国VP建筑设计事务所
项目规模：192 000 m²
建筑面积：123 721 m²
建成时间：2014年
设计周期：11个月

Owner: CITS (Sanya) Investment Development Co., Ltd.
Architectural Design: Valode & Pistre Architecture Design Consulting(Beijing)Co., Ltd.
Project Scale: 192,000 m²
Floor Area: 123,721 m²
Date of Completion: 2014
Cycle of Design: 11 months

第五届·金奖
5th GOLD AWARD

Mixed-use ·综合体·

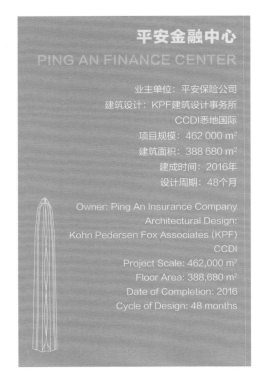

平安金融中心
PING AN FINANCE CENTER

业主单位：平安保险公司
建筑设计：KPF建筑设计事务所
　　　　　CCDI悉地国际
项目规模：462 000 m²
建筑面积：388 680 m²
建成时间：2016年
设计周期：48个月

Owner: Ping An Insurance Company
Architectural Design:
Kohn Pedersen Fox Associates (KPF)
CCDI
Project Scale: 462,000 m²
Floor Area: 388,680 m²
Date of Completion: 2016
Cycle of Design: 48 months

平安金融中心拥有迄今为止世界上最大的不锈钢外立面，作为深圳最高的建筑，它将成为新中央商务区的标志性中心。

The Ping An Finance Center has the largest stainless steel facade in the world till now. As the tallest building in Shenzhen, it will become the iconic center of the new Central Business District.

周大福金融中心
CTF FINANCE CENTER

业主单位：新世界发展有限公司
建筑设计：KPF建筑设计事务所
本地设计单位：广州市设计院
执行建筑师：Leigh & Orange
项目规模：508 000 m²
建筑面积：390 800 m²
建成时间：2016年
设计周期：72个月

Owner: New World Development
Architectural Design: Kohn Pedersen
　　　　　Fox Associates(KPF)
Local Design Institute: Guangzhou
　　　　　Design Institute
Executive Architect: Leigh & Orange
Project Scale: 508,000 m²
Floor Area: 390,800 m²
Date of Completion: 2016
Cycle of Design: 72 months

第五届·金奖
5th GOLD AWARD

Commercial Project ·商办项目·

作为中国第二高的建筑，周大福金融中心通过巧妙的规划设计和垂直体量阶梯变化满足楼层间的各种功能需求，明确办公、住宅、酒店和塔冠的四个过渡点。

As the second tallest building in China, the CTF Finance Center meets various functional needs through skillful planning and design and vertical volume step changes, clarifying the four transition points of office, residence, hotel and tower crown.

第五届·金奖
5th GOLD AWARD

Urban Regeneration ·城市再生·

船厂1862
MIFA 1862

业主单位：上海中船置业有限公司
建筑设计：隈研吾建筑都市设计事务所
项目规模：31 626 m²
建筑面积：26 010 m²
建成时间：2017年
设计周期：29个月

Owner: Shanghai Chime Shipping Property Ltd.
Architectural Design: Kengo Kuma and Associates
Project Scale: 31,626 m²
Floor Area: 26,010 m²
Date of Completion: 2017
Cycle of Design: 29 months

设计团队将建造于1972年的造船厂改造为由剧院、商业中心两个功能组成的综合设施。造船厂为制作巨大尺度的货轮而建，享有独特的尺度感，改建后的船厂依旧保留这种有象征意义的尺度感。

The design team transformed the shipyard built in 1972 into a comprehensive facility consisting of two functions: a theater and a commercial center. MIFA 1862 is built for the production of large-scale freighters to enjoy a unique sense of scale. The re-built shipyard still retains this symbolic sense of scale.

腾讯总部
TENCENT HEADQUARTERS

业主单位：腾讯控股有限公司	Owner: Tencent Holdings Limited.
建筑设计：NBBJ建筑设计事务所	Architectural Design: NBBJ
项目规模：18 700 m²	Project Scale: 18,700 m²
建筑面积：345 570 m²	Floor Area: 345,570 m²
建成时间：2017年	Date of Completion: 2017
设计周期：14个月	Cycle of Design: 14 months

Headquarter · 企业总部 ·

腾讯总部通过"空中街道"将高层建筑所划分的双塔楼衔接到一起，引入社区空间、绿化空间和康体配套空间等连通元素，鼓励腾讯员工在身心和创造力层面的发展。

Tencent headquarters connects the two independent towers through the "sky street", introducing the linking elements such as community space, green space, and recreational space into this project, encouraging Tencent employees to develop their physical and mental health, and creativity.

LCM 置汇旭辉广场
LCM PLAZA

业主单位：旭辉集团&香港置地
建筑设计：Benoy 贝诺
项目规模：340 167 m²
建筑面积：226 667 m²
建成时间：2018年

Owner: CIFI GROUP&Hongkong Land Ltd.
Architectural Design: Benoy
Project Scale: 340,167 m²
Floor Area: 226,667 m²
Date of Completion: 2018

Mixed-use · 综合体 ·

LCM 以"Let's Create More"（让我们创造更多）为理念打造未来城市新地标，设计灵感来自洋泾港的丰富历史资源，提炼出"街，园，节，博，墙"的设计语言，营造一个有历史场所感、开放灵活、优美的综合商业空间。

With the concept of "Let's Create More", LCM creates a new landmark for the future city. Inspired by the rich historical resources of Yangjing Port, the project extracts the design language of "street, garden, festival, blog, wall", creating a historical, convenient, beautiful, and multi-functional commercial space.

万科时代中心 / 北京十里堡
VANKE TIMES CENTER / BEIJING SLP

业主单位：北京万科企业有限公司
建筑改造与室内设计：SHL建筑事务所
项目规模：47 000 m²
建成时间：2018年
设计周期：24个月

Owner: Beijing Vanke Co., Ltd.
Architecture and Interior Design: Schmidt Hammer Lassen Architects
Project Scale: 47,000 m²
Date of Completion: 2018
Cycle of Design: 24 months

Urban Regeneration · 城市再生 ·

项目位于北京繁华的朝阳区十里堡，是一个充满创意的全新城市综合体。融合精品商业、文化办公、大型艺术装置、多功能展览空间和"冥想竹园"于一体，是新型的城市地标。

Located in Beijing's busy Chaoyang District, Vanke Times Center is a new creative urban complex that blends retail shops and offices with grand art installations, a multi-functional exhibition space, and a bamboo meditation garden.

第五届·金奖
5th SILVER AWARD

Mixed-use Complex ·综合体·

武汉天地壹方北馆
NORTH HALL OF WUHAN TIANDI HORIZON

业主单位：瑞安房地产
建筑设计：5+design 五杰建筑设计
华东建筑设计研究总院
建筑面积：74 269 m²
建成时间：2019年
设计周期：60个月

Owner: Shui On Land
Architectural Design: 5+design; ECADI
Floor Area: 74,269 m²
Date of Completion: 2019
Cycle of Design: 60 months

项目致力于保护快速城市发展中的文化景观要素，延续武汉天地的公共绿地，创造出有别于其他商场的绿色公共空间，将景观延伸到城市街道上，形成一幅渐次展开的山水画卷。

Our project draws from the geographical landscape to ensure that the essential elements of Wuhan's rich cultural landscape are protected. It extends the existing park in Wuhantiandi into the mall, thus creating a special public space that other malls usually won't have. As the gardens spill out on to the streetscape, which has formed a slowly unfolding Chinese traditional painting scroll.

水晶大楼—明斯特保险公司总部
CRYSTAL—LVM5 HEADQUARTERS, MÜNSTER

业主单位：德国明斯特州立保险公司
建筑设计：HPP建筑事务所
建筑面积：19 800 m²
建成时间：2014年
设计周期：48个月

Owner: LVM Landwirtschaftlicher Versicherungsverein Münster a. G.
Architectural Design: HPP Architects
Floor Area: 19,800 m²
Date of Completion: 2014
Cycle of Design: 48 months

第四届·银奖
4th SILVER AWARD

Headquarter ·企业总部·

LVM5因其独特的不规则几何形体被称为"水晶大楼"。项目由垂直的18层塔楼、倾斜的外立面和三层三角形基座构成。水晶大楼倾斜的立面向旁边的建筑微微靠拢，并通过位于11层的玻璃廊桥与之相连。外层幕墙通过三角形的网格更好地诠释出建筑的几何形态。

The LVM 5 building was well-known as the 'Crystal Building' due to its distinctive cubism. It includes 18-storey tower with its sloping facades and 3-storey triangular plinth. The Crystal faces the neighboring buildings through its inclined facade planes and connected to them through a glass bridge on the 11th floor. The external facade follows the triangular grid on which the geometry of the building as a whole is based.

第四届·银奖
4th SILVER AWARD

Culture Project ·文化项目·

银川韩美林艺术馆
YINCHUAN HAN MEILIN ART MUSEUM

业主单位：银川市贺兰山岩画管理处
建筑设计：SUNLAY三磊
占地面积：15 866 m²
建筑面积：6 694 m²
建成时间：2015年
设计周期：24个月

Owner: Administrative Office of Helan Mountain Rock Arts in Yinchuan
Architectural Design: SUNLAY
Site Area: 15,866 m²
Floor Area: 6,694 m²
Date of Completion: 2015
Cycle of Design: 24 months

设计灵感源于贺兰山的苍茫雄壮及当地居民传承的因地制宜的所房屋建造方式，利用山势高差，将建筑整体嵌入山体，引入日光与山景。外墙面毛石均就地取材，是目前银川最高的外装毛石砌筑建筑，表现现代艺术与大自然的对话。

The design is inspired by the majesty of Helan Mountain, and the heritage of adjusting measures to local conditions when building houses. The height difference of the mountain is used to embed the building as a whole into the mountain, introducing sunlight and mountain views. The material of the Art Museum facade is ashlar, which utilizes local material in Helan Mountain. The Art Museum is the tallest masonry wall structure in Yinchuan, providing a conversation between modern arts and nature.

郑州普罗理想国艺术文化中心
ZHENGZHOU PULUO IDEAL LAND—ART & CULTURE CENTER

业主单位：郑州普罗房地产开发有限公司
建筑设计：建言建筑 Verse Design
项目规模：12 000 m²
建筑面积：5 100 m²
建成时间：2015年
设计周期：12个月

Owner: Zhengzhou Province Property Development Co., Ltd.
Architectural Design: Verse Design
Project Scale: 12,000 m²
Floor Area: 5,100 m²
Date of Completion: 2015
Cycle of Design: 12 months

Sales Center ·售楼中心·

艺术文化中心是普罗理想国社区的生活缩影与精神象征，功能区域被分别设计成不同的长方形体量，通过墙与连廊进行串连，借助空间的虚实转换形成建筑主体与庭院空间的自然过渡，憧憬未来理想生活。

The Art and Culture Center is the spiritual symbol and the microcosm of life of the Puluo Ideal Land community. Its functional areas are composed of different rectangular buildings, and each building is connected through walls and corridors, forming the natural transition between the main body of the buildings and the courtyard space by means of space transformation to look forward to an ideal life in the future.

Mixed-use ·综合体·

世贸一期是浦东陆家嘴前滩开发的重要部分，东面坐落两栋135米高的甲级办公塔楼，西面为3层高的商业街区。半露天式的裙楼配套齐全，为周围上班族、游客及附近居民提供多层次的休闲空间。

The project is an important part of the development of the large beach in Lujiazui, Pudong. The project has two 135-meter-high Grade A office towers in the east, and a three-story commercial block in the west. The semi-outdoor podium houses are well equipped, providing multi-level people-oriented leisure spaces for surrounding office workers, visiting tourists and nearby residents.

世贸一期
THE NEW BUND WORLD TRADE CENTER (PHASE I)

业主单位：陆家嘴集团
&上海前滩国际商务区投资(集团)有限公司
建筑设计：Benoy 贝诺
项目规模：95 000 m²
建筑面积：140 000 m²
建成时间：2017年
设计周期：24个月

Owner: Lujiazui Group
&Shanghai New Bund International Business District Investment (Group) Co., Ltd.
Architectural Design: Benoy
Project Scale: 95,000 m²
Floor Area: 140,000 m²
Date of Completion: 2017
Cycle of Design: 24 months

枣庄市文体中心体育场
ZAOZHUANG SPORT AND CULTURE CENTER STADIUM

业主单位：枣庄金声文化产业发展有限公司
建筑设计：上海联创设计集团股份有限公司
项目规模：64 582 m²
建筑面积：56 000 m²
建成时间：2017年
设计周期：24个月

Owner: Zaozhuang JinSheng Culture Development Co., Ltd.
Architectural Design: Shanghai United Design Group Co., Ltd.
Project Scale: 64,582 m²
Floor Area: 56,000 m²
Date of Completion: 2017
Cycle of Design: 24 months

Culture Project ·文化项目·

项目坐落于枣庄文化体育公园的核心区域，体育场曲线的立面设计灵感来源于"运河之乡"的流水与极具中国传统文化代表性的中国灯笼，不同曲度的波浪线形结构围合形成整体立面造型。

The project is in the heart of Zaozhuang Cultural and Sports Park. The curved facade of the stadium is inspired by the flowing water of the "Home of the Canal" and the Chinese lantern, with wave lines of different curvatures that form the overall facade.

Mixed-use Complex ·综合体·

项目位于上海核心枢纽"大虹桥"区域中,各个建筑单体以"花心、花瓣、绿叶"的形式进行有机排列形成"向阳花"。"花心"与国展中心轴心对位,遥相呼应,构成区域内最具辨识性的建筑群。

Hongqiao World Center lies in the Hongqiao Business District that is the pivot of Shanghai. The buildings are organized coherently to direct circulations, presenting an analogy of sunflower with the style of stamen, petal and leaf. The center of flower is aligned with its counterpart of NECC, attaining the most identifiable building clusters.

虹桥世界中心
HONGQIAO WORLD CENTER

业主单位:绿地控股集团&融信集团
规划设计:UA尤安设计
建筑设计:UA尤安设计(花心花瓣南叶北叶东叶)
项目规模:184 292 m²
建筑面积:821 812 m²
建成时间:2018年

Owner: GREENLAND HOLDING GROUP
&RonShine Group
Plannning Design: UA
Architecture Design: UA(Flower-Shaped/Petal-Shaped/
Southern/Northern/Eastern buildings)
Project Scale: 184,292 m²
Floor Area: 821,812 m²
Date of Completion: 2018

曼海姆商学院研究与会议中心
MBS STUDY AND CONFERENCE CENTRE

业主单位:巴登-维腾贝格资产与工程管理公司
&曼海姆和海德堡办事处
建筑设计:德国施耐德+舒马赫规划设计有限公司
项目规模:1 700 m²
建筑面积:1 390 m²
建成时间:2017年
设计周期:21个月

Owner: Vermögen und Bau Baden-Württemberg
&Amt Mannheim
Architectural Design: schneider+schumacher
Planungsgesellschaft mbH
Project Scale: 1,700 m²
Floor Area: 1,390 m²
Date of Completion: 2017
Cycle of Design: 21 months

Culture Project ·文化项目·

曼海姆城堡不仅是一个历史文物,更是一个繁荣的合奏。本项目利用老城堡建造全新的教学空间,将城堡煤窑变身为露天景观,极富前瞻性的设计理念也让曼海姆商学院完美地嵌入历史建筑中。

Mannheim Castle is not only a historical relic, but also a prosperous ensemble. The design uses the old castle to build a new teaching space. Together with the historic castle, this complex-cut into the existing garden-creates a distinctive new component that symbolizes the preservation of, and respect for tradition, as well as the importance of innovation and scientific curiosity.

Urban Regeneration ·城市再生·

长春水文化生态园将具有80年历史的水厂改造为文化艺术社区,以景观思维统筹规划建筑、景观、艺术特性,形成慢行系统、原址动植物生态系统和水生态自净化系统,突出生态园区的功能属性和人文特色。

Changchun Culture of Water Ecology Park transforms an 80-year-old water plant into a cultural and artistic community. The planning, architecture, landscape, and art installations are integrated with the landscape thinking. The landscape design is systematic, thereby forming a chronic system, an original site animal and plant ecosystem a water ecological self-purification system, and highlighting the function and humanistic feature.

长春水文化生态园
CHANGCHUN CULTURE OF WATER ECOLOGY PARK

业主单位:长春建委&长春城投
建筑设计:水石设计
项目规模:340 000 m²
建筑面积:50 000 m²
建成时间:2018年
设计周期:24个月

Owner:
Changchun Urban and Rural Construction Committee
&Changchun Construction Investment Co., Ltd.
Architectural Design: SHUISHI
Project Scale: 340,000 m²
Floor Area: 50,000 m²
Date of Completion: 2018
Cycle of Design: 24 months

荃湾体育馆
TSUEN WAN SPORTS CENTRE

业主单位：新世界发展有限公司
建筑设计：LWK + PARTNERS
项目规模：12 000 m²
建成时间：2018年
设计周期：60个月

Owner: New World Development Company Limited.
Architectural Design: LWK + PARTNERS
Project Scale: 12,000 m²
Date of Completion: 2018
Cycle of Design: 60 months

第五届·银奖
5th SILVER AWARD

Culture Project · 文化项目 ·

项目定位为国际体坛盛事的新家园，建筑外立面以简单几何形为主轴，流动空间视线开阔，自然转换活动空间，为社区增值的同时满足居民的使用需求，是一个真正能够容纳所有使用者的公共王国。

The project is positioned as a new home for international sport events. The external facade takes simple geometry as the main axis, and the flow space has a wide view, which naturally transforms the activity space. It is aesthetically pleasing and functionally adequate and satisfies both visual and user experience. It is a public realm that can truly accommodate all users.

第五届·银奖
5th SILVER AWARD

盛港综合及社区医院
SENGKANG GENERAL AND COMMUNITY HOSPITAL

业主单位：卫生部
建筑设计：DP 建筑师事务所
项目规模：288 000 m²
建成时间：2018年
设计周期：18个月

Owner: Ministry of Health
Architectural Design: DP Architects Pte Ltd.
Project Scale: 288,000 m²
Date of Completion: 2018
Cycle of Design: 18 months

Healthcare Project · 医养项目 ·

项目设计愿景是打造新加坡东北部最顶尖的医疗中心，打破人们对传统医疗空间的认知，促进整体资源的高效利用，营造环境宜居、健康全面的医疗、社区空间。

The design of the Sengkang General and Community Hospital aims to be the best regional healthcare provider for the North-East region of Singapore. It aims to create a leading integrated medical facility to promote efficient use of resources. It wishes to promote healing for everyone, integrating into the heartland of Sengkang and creating spaces and places for the community.

第五届·银奖
5th SILVER AWARD

亚马逊星球
AMAZON SPHERES

业主单位：亚马逊公司
建筑设计：NBBJ建筑设计事务所
摄　　影：Benjamin Benschneider
项目规模：7 242 m²
建筑面积：6 500 m²
建成时间：2018年
设计周期：22个月

Owner: Amazon.com, Inc.
Architectural Design: NBBJ
Photographer: Benjamin Benschneider
Project Scale: 7,242 m²
Floor Area: 6,500 m²
Date of Completion: 2018
Cycle of Design: 22 months

Headquarter · 企业总部 ·

亚马逊星球为人与植物的共生提供了独一无二的场所，不再以隔间或办公桌的形式呈现办公场所，取而代之的是一个生机勃勃、洒满阳光的园艺空间，这里共有 400 种植物，总量超过 40 000 株。

The Amazon Spheres provide a unique workplace where both people and plants thrive. They were conceived as an alternative workspace with no rooms or desks, an invigorating, nature-filled space housing a horticultural collection of more than 40,000 plants from 400 species.

02 居 住 项 目 RESIDENTIAL PROJECT

第一届金奖

1st GOLD AWARD

合肥合院别墅
HEFEI COURTYARD VILLA

南京九间堂
JIU JIAN TANG (MANDARIN PALACE), NANJING

第二届金奖

2nd GOLD AWARD

华侨城苏河湾东区 88 号院
OCT SUHE CREEK EAST NO.88

北京 中粮 瑞府
COFCO THE GARDEN OF EDEN, BEIJING

厦门建发五缘湾央座
XIAMEN C&D WUYUAN BAY CENTRAL MANSION

第三届金奖

3rd GOLD AWARD

苏州中航樾园
AVIC SUZHOU YUEYUAN

万科良渚文化村未来城二期桂语里
VANKE LIANGZHU NEW TOWN FUTURE LIFE PHASE II – GUIYULI

北京壹号院
BEIJING ONE SINO PARK

合肥旭辉铂悦庐州府
HEFEI CIFI PLATINUM WYATT LUCHOWFU

尚东柏悦府
THE LANDMARK

第四届金奖 / 4th GOLD AWARD

长城脚下饮马川——拾得大地幸福实践区
YIN MA CHUAN OF THE GREAT WALL—FINDING HAPPINESS PRACTICE AREA

武汉碧桂园晴川府
WUHANG COUNTRY GARDEN QINGCHUAN MANSION

第五届金奖 / 5th GOLD AWARD

重庆龙湖舜山府
CHONGQING LONGFOR SHUNSHAN MANSION

君山生活美学馆
JUNSHAN CULTURAL CENTER

远洋集团·武汉东方境世界观
SINO-OCEAN WUHAN ORIENTAL WORLD VIEW

第四届银奖 / 4th SILVER AWARD

绿地黄浦滨江
GREENLAND HUANGPU CENTER

凝白之间
SCULPTURE WHITE

瀚海晴宇
HANHAI LUXURY CONDOMINIUMS

南宁万科金域中央
VANKE NANNING JINYU CENTRAL

第五届银奖 / 5th SILVER AWARD

上海仁恒公园世纪
SHANGHAI YANLORD PARK CENTURY

建业濮阳世和府示范区
JIANYE PUYANG SHIHE MANSION DEMONSTRATION AREA

圣多马士八号公寓
8 SAINT THOMAS

泰禾·青云小镇
TAHOE QINGYUN TOWN

上海 融信 世纪江湾
RONSHINE CENTURY SUMMIT, SHANGHAI

合肥合院别墅
HEFEI COURTYARD VILLA

业主单位：安徽盛通置业有限公司
建筑设计：维思平建筑设计
项目规模：13 333 m²
建筑面积：28 408 m²
建成时间：2014年
设计周期：3个月

Owner: Anhui Shengtong Property Co., Ltd.
Architectural Design: WSP ARCHITECTS
Project Scale: 13,333 m²
Floor Area: 28,408 m²
Date of Completion: 2014
Cycle of Design: 3 months

第一届·金奖
1st GOLD AWARD

Residential Project · 居住项目 ·

本项目用现代的表现手法演绎传统，自成院落的形式实现风景与建筑的统一，为家庭提供更为丰富的生活场景。各院落并置串联形成院落群轴，满足家庭乐趣的同时，业主也能从邻里交往中获得社会属性的满足感。

The original intention of this project is to realize the tradition with modern expression methods. The self-contained courtyard realizes the unity of scenery and architecture and provides a richer life scene for the family. The courtyards are constitutes the life of the neighborhood, juxtaposed in series to form the axis, which not only satisfies the family fun, but also obtains the satisfaction of social attributes from the neighborhood interaction.

南京九间堂
JIU JIAN TANG(MANDARIN PALACE), NANJING

业主单位：江苏证大商业文化发展有限公司
建筑设计：许李严建筑师事务有限公司
项目规模：256 466.66 m²
建筑面积：51 614 m²
建成时间：2016年

Owner: Jiangsu Zendai Commercial Culture Development Co., Ltd.
Architectural Design: Rocco Design Architects Ltd.
Project Scale: 256,466.66 m²
Floor Area: 51,614 m²
Date of Completion: 2016

第一届·金奖
1st GOLD AWARD

Residential Project · 居住项目 ·

项目环抱康厚水库，背靠将军山，区位优势造就有机布局，让别墅依山而建，绕湖而列，坐拥最丰富的天然景观。现代化的坡顶设计呈现中式大宅的气派，令建筑与景致和谐结合，是文化与自然的结晶。

The project surrounds the Kanghou Reservoir and is backed by Jiangjun Mountain. The geographical advantage creates an organic layout, allowing the villas to be built on the mountain and around the lake, with the most abundant natural landscape. The modern sloped roof design presents the style of a Chinese-style mansion, harmoniously combining architecture and scenery, and is the crystallization of culture and nature.

第二届·金奖
2nd GOLD AWARD

华侨城苏河湾东区 88 号院
OCT SUHE CREEK EAST NO.88

业主单位：华侨城（上海）置地有限公司
规划设计：Foster + Partners
建筑设计：Foster+Partners
KOKAIstudios
华东都市建筑设计研究总院
建筑面积：234 091 m²
建成时间：2014年
设计周期：36个月

Owner: OCT Land (Shanghai) Investment Ltd.
Planning Design: Foster + Partners
Architectural Design: Foster+Partners
KOKAIstudios
East China Urban Architectural Design & Research Institute
Floor Area: 234,091 m²
Date of Completion: 2014
Cycle of Design: 36 months

Residential Project · 居住项目 ·

项目是集商业、酒店、住宅、办公、历史建筑于一体的超高层商办综合楼，中低层区为公寓式办公区，高区层为BVLGARI酒店的客房层，配有意大利餐厅和酒吧的顶部会所让游客尽享黄浦江景。

The project is a super high-rise commercial and office complex integrating commerce, hotel, residence, office and historical buildings. The middle and low floors are apartment-style office areas, and the upper floors are the guest rooms of the BVLGARI hotel, with an Italian restaurant and bar on the top. The clubhouse allows visitors to enjoy the view of Huangpu River.

Residential Project · 居住项目 ·

北京 中粮 瑞府
COFCO THE GARDEN OF EDEN, BEIJING

业主单位：中粮大悦城集团
建筑设计：JUND骏地设计
项目规模：69 700 m²
建筑面积：76 700 m²
建成时间：2015年
设计周期：14个月

Owner: GRANDJOY
Architectural Design: JUND
Project Scale: 69,700 m²
Floor Area: 76,700 m²
Date of Completion: 2015
Cycle of Design: 14 months

中国文化历来讲究人与自然的和谐共生，多以庭院承载二者关系。因而项目在空间设计中塑造了形态迥异的若干庭院，与功能空间交织融合渗透到建筑群中，呈现出曲径通幽、步移景异的意境。

Chinese culture has always emphasized the harmonious symbiosis between human and nature, and the relationship between the two is carried by courtyards. Therefore, the project has created several courtyards of different shapes in the space design, intertwined with the functional space and penetrated the building complex, presenting the artistic conception of winding paths and different scenery.

厦门建发五缘湾央座
XIAMEN C&D WUYUAN BAY CENTRAL MANSION

业主单位：建发集团
建筑设计：水石设计
项目规模：42 444 m²
建筑面积：42 444 m²
建成时间：2015年
设计周期：21个月

Owner: C&D Group
Architectural Design: SHUISHI
Project Scale: 42,444 m²
Floor Area: 42,444 m²
Date of Completion: 2015
Cycle of Design: 21 months

Residential Project · 居住项目 ·

项目基于高端城市社区定位，力求打造出建筑美学与生活品质完美结合的豪宅作品。得天独厚的内海资源为高端客户打造奢华大平层海景住宅，立面整体造型灵感取自游艇元素，追求动感的艺术境界。

Based on the positioning of high-end urban community, this project strives to create a luxury real estate with the perfect combination of architectural aesthetics and quality of life. The blessed inland sea resources create luxury large flat-level sea view residences for high-end customers. The design is inspired by yacht elements, and the effect presents a sense of fluency and simplicity.

Residential Project · 居住项目 ·

苏州中航樾园
AVIC SUZHOU YUEYUAN

业主单位：苏航置业有限公司
建筑设计：北京市建筑设计研究院有限公司·王戈工作室
项目规模：180 000 m²
建筑面积：180 000 m²
建成时间：2016年
设计周期：6个月

Owner: Suhangzhiye Co., Ltd.
Architectural Design: BIAD · BOA
Project Scale: 180,000 m²
Floor Area: 180,000 m²
Date of Completion: 2016
Cycle of Design: 6 months

项目位于苏州木渎镇，设计师借鉴苏州传统风貌，将老房子与丝网印刷玻璃搭配，以流动的体量包裹外围，外立面追求自然侵蚀的粗糙效果，内院则构造多层次的流动空间，演绎现代韵味的"苏式园林"。

The project is in Mudu Town. Designers choose to respect and learn from Suzhou's traditional style. The old house is matched with gradient screen-printed glass, and the exterior is wrapped in a flowing volume, deducing a modern "Su-style garden".

万科良渚文化村未来城二期桂语里
VANKE LIANGZHU NEW TOWN FUTURE LIFE PHASE II – GUIYULI

业主单位：万科集团
建筑设计：AAI国际建筑师事务所
项目规模：23 000 m²
建筑面积：22 500 m²
建成时间：2016年
设计周期：5个月

Owner: Vanke
Architectural Design: ALLIED ARCHITECTS INTERNATIONAL
Project Scale: 23,000 m²
Floor Area: 22,500 m²
Date of Completion: 2016
Cycle of Design: 5 months

Residential Project ·居住项目·

桂语里借鉴杭州的老式坊巷，采用传统民居宅第类建筑的构成模式打造一个"骨子里的江南"。其规划布局通过"多进 + 多跨 + 园宅组合"模式，打造小而美的"乐高"微墅。

The mini villas, also known as Guiyuli, display a classical 'Jiangnan' style by using the traditional structure of the old lanes and alleys in Hangzhou. The design aims to build a small but beautiful "Lego" villa with the model of "Multi-entry+Multi-span+Garden".

北京壹号院
BEIJING ONE SINO PARK

业主单位：融创中国
建筑设计：goa大象设计
建筑面积：89 000 m²
建成时间：2017年
设计时间：48个月

Owner: SUNAC
Architectural Design: GOA
Floor Area: 89,000 m²
Date of Completion: 2017
Cycle of Design: 48 months

Residential Project ·居住项目·

项目位于北京核心地带，力求突破中国奢华住宅的形式窠臼，营造一处北京城内的现代精品社区。设计师运用富有现代感的建筑材料和建构技艺塑造整体形象，探索中国高端居住的全新模式。

Beijing One Sino Park is located in the core area of Beijing and strives to break through the formal stereotypes of Chinese luxury residences and creates a modern boutique community in Beijing. The designers use modern building materials and construction techniques to shape the overall image and explore a new model of high-end residential in China.

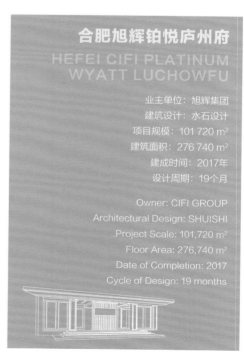

合肥旭辉铂悦庐州府
HEFEI CIFI PLATINUM WYATT LUCHOWFU

业主单位：旭辉集团
建筑设计：水石设计
项目规模：101 720 m²
建筑面积：276 740 m²
建成时间：2017年
设计周期：19个月

Owner: CIFI GROUP
Architectural Design: SHUISHI
Project Scale: 101,720 m²
Floor Area: 276,740 m²
Date of Completion: 2017
Cycle of Design: 19 months

Residential Project ·居住项目·

项目为合肥滨湖高端生态居住区，总体布局舒适宜居，采用新亚洲的建筑风格将亚洲元素植入建筑语系，传统意境和现代风格对称运用隐喻中国传统，是为精英人士提高居住环境而打造的高端居住社区。

The project is a high-end ecological residential area built for the elites to improve their living condition. The architecture adopts the new Asian architectural style. The traditional artistic conception and modern style are used symmetrically.

尚东柏悦府
THE LANDMARK

业主单位：尚东控股集团	Owner: Top East Investment Holding Group
建筑设计：汉森伯盛国际设计集团	Architectural Design: Shing & Partners Design Group
项目规模：14 300 m²	Project Scale: 14,300 m²
建筑面积：96 000 m²	Floor Area: 96,000 m²
建成时间：2017年	Date of Completion: 2017
设计周期：108个月	Cycle of Design: 108 months

Residential Project · 居住项目 ·

尚东柏悦府位于广州珠江新城，是一座高198米的超高层建筑，为目前广州最高的纯住宅项目。其有机曲面回应了广州塔及珠江的现存肌理，以建筑设计发挥地脉价值。"三塔合一"的布局打破了传统思维，重塑城芯豪宅的极致体验。

Located in Zhujiang New Town, The Landmark is a tall building with the height of 198m, which is the tallest residential project within the city. With its organic appearance, the architecture responds to the existing texture composed by Canton Tower and the Pearl River, maximizing the site value from the architectural design perspective. Its iconic Three-in-One layout breaks conventional mindset, and The Landmark is redefining the ultimate experience of high-end lifestyles.

长城脚下饮马川
——拾得大地幸福实践区
YIN MA CHUAN OF THE GREAT WALL—FINDING HAPPINESS PRACTICE AREA

业主单位：拾得大地幸福集团有限公司	
建筑设计：澳大利亚IAPA设计顾问有限公司	
项目规模：350 000 m²	
建筑面积：85 768 m²	
建成时间：2017年	
设计周期：8个月	

Owner: The Mother Earth Happiness Group
Architectural Design: IAPA Pty. Ltd.
Project Scale: 350,000 m²
Floor Area: 85,768 m²
Date of Completion: 2017
Cycle of Design: 8 months

Demonstration Area · 示范区 ·

项目的建立旨在让城市人逃离城市的压力，拾回清净，回归田园，放松身心。本项目关注自然环境与村落复兴，以生态艺术实现原生资源的再创造，致力打造国内第一个大地艺术生态公园，建立"艺术生态涵养生活区"。

The establishment of the project aims to allow people to escape the stress of the city and return to a clean, natural environment where they can relax and nurture their soul. The project pays attention to the natural environment and the revival of the village and realize the re-creation of original resources with ecological art, as well as committes to creating the first land art ecological park in China and establishes an "art ecological conservation living area".

武汉碧桂园晴川府
WUHANG COUNTRY GARDEN QINGCHUAN MANSION

业主单位：碧桂园集团
建筑设计：水石设计
项目规模：31 610 m²
建筑面积：170 140 m²
建成时间：2017年
设计周期：8个月

Owner: Country Garden
Architectural Design: SHUISHI
Project Scale: 31,610 m²
Floor Area: 170,140 m²
Date of Completion: 2017
Cycle of Design: 8 months

Residential Project · 居住项目 ·

"晴川府"为碧桂园在武汉的第一个地标性、超高层高端居住社区，项目用地一线临江，景观资源得天独厚，将醇厚的历史底蕴与城市核心活力完美融合，周边配套设施成熟，区域价值无出其右。

The Qingchuan Mansion, as the first landmark established by Country Garden in Wuhan, is a high-end residential community. Facing the river, the project has a unique landscape resource. The project perfectly integrates the pro-found historical background with the core vitality of modern cities and has a high regional value.

重庆龙湖舜山府
CHONGQING LONGFOR SHUNSHAN MANSION

业主单位：重庆龙湖	Owner: Longfor Chongqing
建筑设计：成执设计	Architectural Design: Challenge Design
项目规模：79 287 m²	Project Scale: 79,287 m²
建筑面积：283 660.66 m²	Floor Area: 283,660.66 m²
建成时间：2018年	Date of Completion: 2018
设计周期：29个月	Cycle of Design: 29 months

5th GOLD AWARD

Residential Project · 居住项目 ·

龙湖舜山府通过一环、一带、多线轴与照母山紧密连接，形成独一无二的高价值都市场所。山水之城流线型的建筑形态带来一种"矛盾"的美，创造高品质景观社区与自然环境的和谐对话。

Chongqing Longfor Shunshan Mansion is closely linked to Zhaomu Mountain through one ring, one belt, and multiple spools, forming a unique high-value urban place. The streamlined architectural form of the city brings a kind of "contradictory" beauty, creating a harmonious dialogue between high-quality landscape communities and the natural environments.

5th GOLD AWARD

Sales Center · 售楼中心 ·

君山生活美学馆
JUNSHAN CULTURAL CENTER

业主单位：阳光城北京区域公司	Owner: Yango Beijing
建筑设计：如恩设计研究室	Architectural Design: Neri & Hu Design and Research Office
项目规模：4 277 m²	Project Scale: 4,277 m²
建筑面积：4 277 m²	Floor Area: 4,277 m²
建成时间：2018年	Date of Completion: 2018

让建筑融于自然之中，让人与自然纯粹地结合是君山美墅生活美学馆建造的初衷。设计师将北京当地传统建筑手法和现代建筑语言进行糅合汇聚，让生活美学馆成为倡导生活美学的地标性体验馆。

The original intention of Junshan Aesthetics Museum is to let the architecture blend in with nature, integrate people and nature as well. It combines the traditional local architectural techniques in Beijing with modern architectural language to make it a landmark for people to experience the aesthetics of life.

远洋集团·武汉东方境世界观
SINO-OCEAN WUHAN ORIENTAL WORLD VIEW

业主单位：远洋集团	Owner: SINO-OCEAN Land
建筑设计：TIANHUA天华	Architectural Design: TIANHUA
项目规模：4 337 m²	Project Scale: 4,337 m²
建筑面积：4 337 m²	Floor Area: 4,337 m²
建成时间：2018年	Date of Completion: 2018
设计周期：5个月	Cycle of Design: 5 months

5th GOLD AWARD

Sales Center · 售楼中心 ·

项目力图打造区域特色的纪念性城市场景，形成具有公共属性的建筑聚落，以归元禅寺为起点，采用递进式院落布局围合而成内向型的院落空间，用现代手法诠释江南特色，回应当下"城市更新"的社会主题。

The project aims to be a monumental urban landscape with regional characteristics, which will become a collection of buildings for public activities and experience. The planning starts from the Guiyuan Temple, adopting a progressive courtyard layout unconventional. Modern design techniques are used to showcase Jiangnan Style and responses to today's hot topic of "urban renewal".

绿地黄浦滨江
GREENLAND HUANGPU CENTER

业主单位：绿地控股
建筑设计：TIANHUA天华
　　　　　联合设计：骏地设计
项目规模：28 179 m²
建筑面积：28 179 m²
建成时间：2016年
设计周期：23个月

Owner: Greenland
Architectural Design: TIANHUA
　　　　　Joint Design: JUND
Project Scale: 28,179 m²
Floor Area: 28,179 m²
Date of Completion: 2016
Cycle of Design: 23 months

第四届·银奖
4th SILVER AWARD

Residential Project · 居住项目 ·

项目位于上海世博滨江板块核心地段，旨在最大化利用江景自然资源，打造外有"江"内有"院"的规划格局，结合工业4.0时代精工技艺与美学理念，为住户量身定制"推窗见江景"的高品质生活体验。

Greenland Huangpu Center, situated in the heart of the Shanghai World Expo, aims to maximally utilize the natural riverside resources to create a location with "a river outside" and "a courtyard inside", and to provide a tailor-made high-quality living experience of "enjoying the river view upon opening the window" for every resident by combining the great craftsmanship of the Industry 4.0 era with the aesthetics.

凝白之间
SCULPTURE WHITE

业主单位：朋澄-广美建材
建筑设计：黄静文室内设计有限公司
项目规模：535 m²
建筑面积：535 m²
建成时间：2016年
设计周期：2个月

Owner: Perng Cherng-G.M. Building Materials Co., Ltd.
Architectural Design: D.H.I.A international design Co., Ltd.
Project Scale: 535 m²
Floor Area: 535 m²
Date of Completion: 2016
Cycle of Design: 2 months

第四届·银奖
4th SILVER AWARD

Private House · 私 宅 ·

设计者将建筑物塑造成一个白色的雕塑品，让不规则的窗户开口如音符跳跃般陈列，保护室内隐私的同时将通风、光照一并纳入设计思考。寻一片干净乐土，从简尽享自然之美，让居所与自然融为一体。

The designers define the building as a white architectural sculpture, with irregular window shapes as vivacious musical notes, well-attuned to the artful purpose of the windows. The size of each window is perfectly scaled for taking tenants' privacy, natural ventilation, and sunlight into considerations. It is designed to suit the tenants to create a calm, peaceful and harmony space for the tenants.

第四届·银奖
4th SILVER AWARD

Residential Project · 居住项目 ·

瀚海晴宇
HANHAI LUXURY CONDOMINIUMS

业主单位：瀚海大观地产
建筑设计：双栖弧建筑事务所
　　　　　筑弧建筑事务所
项目规模：65 333 m²
建筑面积：225 000 m²
建成时间：2017年
设计周期：17个月

Owner: HANHAI DAGUAN REAL ESTATE
Architectural Design: amphibianArc
　　　　　archimorphic
Project Scale: 65,333 m²
Floor Area: 225,000 m²
Date of Completion: 2017
Cycle of Design: 17 months

瀚海晴宇位于郑州市郑东新区，打造高容积率的空中别墅，创新设计和经济效益的完美结合，造就豪宅标杆项目。

located in Zhengdong district, Zhengzhou City, Hanhai Luxury Condominiums create the Sky villas with high FAR, innovative design, and economic benefits, making this project a benchmark for luxury residences.

Demonstration Area · 示范区 ·

南宁万科金域中央
VANKE NANNING JINYU CENTRAL

业主单位：万科集团
建筑设计：汇张思建筑设计咨询（上海）有限公司
项目规模：117 726 m²
建筑面积：631 631 m²
建成时间：2017年
设计周期：25个月

Owner: Vanke
Architectural Design: HZS
Project Scale: 117,726 m²
Floor Area: 631,631 m²
Date of Completion: 2017
Cycle of Design: 25 months

项目坐落于历史悠久的文化古城——南宁。南宁半城山水半城楼，绿城美誉天下闻名。设计结合地域特色，以充满韵律的现代时尚手法打造一个简约而不简单，视觉冲击力大的城市中央大花园。

This project is in the historic cultural city-Nanning, known as green city, described as a poet saying: where there is river and mountain, awhere the city is hidden behind. The design has combined the local character and rhythmical modern fashional method, to make a design contracted but not simple, with giant visual impression, a grand garden in city center.

上海仁恒公园世纪
SHANGHAI YANLORD PARK CENTURY

业主单位：上海仁恒置地
建筑设计：杰作建筑设计咨询（上海）有限公司
项目规模：55 775.7 m²
建筑面积：210 263.85 m²
建成时间：2016年
设计周期：8个月

Owner: Yanlord Land
Architectural Design: TAL the architects. Co., Ltd.
Project Scale: 55,775.7 m²
Floor Area: 210,263.85 m²
Date of Completion: 2016
Cycle of Design: 8 months

Residential Project · 居住项目 ·

仁恒公园世纪秉承"立体景观住宅景观理念"引入生态会所和阳光地库，突出营造富有活力的生态住区，将社区内部绿化景观向城市辐射，打造出一个生态、健康、景观相结合的"都市高尚国际住宅区"。

Yanlord Park Century adheres to the landscape concept of 3D landscape residential area to introduce ecological clubs and sunny basements into this project, highlighting the creation of vibrant ecological residential area, radiating the green landscape within the community to the city, creating a urban noble international residential district combined with ecology, health and landscape.

Demonstration Area · 示范区 ·

建业濮阳世和府示范区
JIANYE PUYANG SHIHE MANSION DEMONSTRATION AREA

业主单位：建业住宅集团（中国）有限公司
建筑设计：岭界建筑设计咨询（上海）有限公司
占地面积：3 430 m²
建筑面积：1 750 m²
建成时间：2018年
设计周期：4个月

Owner: Central China Real Estate Group(China)Company Limited
Architectural Design: VVS Design
Site Area: 3,430 m²
Floor Area: 1,750 m²
Date of Completion: 2018
Cycle of Design: 4 months

项目旨在打造一个兼具现代艺术和古典文化的区域名片，塑造建筑和景观序列起承转合的空间，使时间与空间、建筑与景观和谐共生，在喧嚣的都市中展现一种隐世近水的意境和情怀。

The design intent is to create a regional architectural name card with contemporary and ancient culture and artistic temperament. It shapes the spatial display of the sequence of architecture and landscape. The goal is to create a work of art where time and space, architecture, and landscape harmonious coexist. This project displaces the leisure and coziness and finds a secluded feeling in the hustle and bustle of the city.

圣多马士八号公寓
8 SAINT THOMAS

业主单位：万国公司
建筑设计：DP建筑师事务所
项目规模：35 770 m²
建筑面积：25 884 m²
建成时间：2018年
设计周期：24个月

Owner: Bukit Sembawang View Pte Ltd.
Architectural Design: DP Architects Pte Ltd.
Project Scale: 35,770 m²
Floor Area: 25,884 m²
Date of Completion: 2018
Cycle of Design: 24 months

5th SILVER AWARD

Residential Project · 居住项目 ·

项目拥有250个单位的豪华住宅，凭借对当地文化的理解，设计师创造出极具现代感和实用性的奢华家园，为住户打造与大自然紧密交流的空间，呈现圣多马士八号独特优雅的美感个性。

St. Thomas 8 comprises of 250 freehold units. With a deep understanding of the local culture, DP Architects creates a modern home with a stylish and practical look. A dynamic sense of rhythm on the facade accentuates an exquisite character and quality of a home, while adding a sleek and elegant building to the lush skyline of Orchard Road.

5th SILVER AWARD

Sales Center · 售楼中心 ·

泰禾·青云小镇
TAHOE QINGYUN TOWN

业主单位：泰禾集团
建筑设计：TIANHUA天华
项目规模：3 000 m²
建筑面积：3 000 m²
建成时间：2018年
设计周期：12个月

Owner: Tahoe Group
Architectural Design: TIANHUA Architecture Planning & Engineering Ltd.
Project Scale: 3,000 m²
Floor Area: 3,000 m²
Date of Completion: 2018
Cycle of Design: 12 months

项目地处福州永泰，自南宋以来积累了近千年的人文脉络。结合宋代儒禅文化，设计建造集休闲、疗养、居住、商业于一体的山水文化小镇，让山水资源与居住价值完美融合，形成对传统气韵的现代表达。

Tahoe Qingyun Town is in Yongtai County, Fuzhou. The region has been famous for its culture and local customs. Based on its natural environment, the architects aim to create a small town that encompasses leisure, wellness, living and commercial functions altogether by adopting the traditional Zen culture of the Song Dynasty. It allows a perfect combination of the unique landscapes and the value as residences, to provide a modern expression of the traditional Chinese style.

上海 融信 世纪江湾
RONSHINE CENTURY SUMMIT, SHANGHAI

业主单位：融信集团
建筑设计：JUND骏地设计
项目规模：39,06 m²
建筑面积：88 454 m²
建成时间：2019年
设计周期：12个月

Owner: Ronshine Group
Architectural Design: JUND
Project Scale: 39,806 m²
Floor Area: 88,454 m²
Date of Completion: 2019
Cycle of Design: 12 months

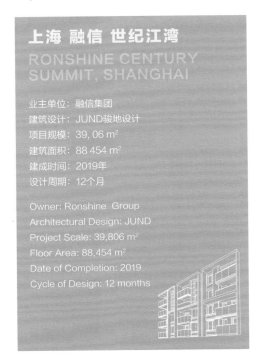

5th SILVER AWARD

Residential Project · 居住项目 ·

设计使用框架结构区分出不同的居住单元，立面承担"院墙"的功能，以空间划分界限，建立私属的领域感。立面设计除了保证整体效果，也通过大量细节提升近景与特写的精致度，提供更优化的视觉感受。

The design uses a frame structure to distinguish different residential units, and the facade assumes the function of a "court wall", divides boundaries with space, and establishes a sense of private domain. The facade design not only guarantees the overall effect, but also enhances the exquisiteness of close-up and close-up through many details, providing a more optimized visual experience.

03 景观设计 LANDSCAPE DESIGN

第一届金奖

1st GOLD AWARD

深圳华侨城欢乐海岸
OCT BAY

黄山雨润涵月楼
HUANGSHAN YURUN HANYUELOU VILLA RESORT

上海万科翡翠滨江
SHANGHAI VANKE EMERALD RIVERSIDE

第二届金奖

2nd GOLD AWARD

重庆东原 1891 印长江
CHONGQING DONGYUAN 1891

福州万科城
FUZHOU VANKE CITY

万科云间传奇
LEGEND OF CLOUDS

第三届金奖

3rd GOLD AWARD

莫奈大道 2.0
MONET AVENUE 2.0 AT VICTORIA GARDENS

北辰旭辉壹号院
CIFI CHATEAU

万科天空之城
UNI-CITY

远洋红岸澜山别墅庭院工程
SINO-OCEAN HONGANLANSHAN VILLA COURTYARD

外滩 SOHO
BUND SOHO

第四届金奖

4th GOLD AWARD

重庆龙湖舜山府展示区
CHONGQING LONGFOR SHUNSHAN MANSION SALES CENTER

福州万科仓前九里项目
FUZHOU VANKE CANGQIANJIULI PROJECT

龙湖新江与城清晖岸
LONGFOR NEW RIVER AND CITY QINGHUI SHORE

龙湖旭辉春江悦茗
LONGFOR CHUNJIANG CENTRAL

● **第五届金奖**
5th GOLD AWARD

泸州·顺成和庄园
LUZHOU SHUNCHENGHE MANOR

音昱水中天
SANGHA BY OCTAVE

贵阳旭辉·观云
SKYLINE

杭州萧山 前湾国际社区·无界公园
XIAOSHAN HANGZHOU BAY UNBOUNDED PARK

宁波万科芝士公园
VANKE CHEESE PARK IN NINGBO

万科·桂语东方
LAUREL ORIENTAL · VANKE

● **第四届银奖**
4th SILVER AWARD

中洲控股金融中心
SCC

北京融创·壹号院
SUNAC ONE SINO PARK IN BEIJING

承德金山岭洛嘉山精灵松塔乐园
CHENGDE JINSHANLING V-ONDERLAND WITH MONTAIN ELF AND PINE CONE

世茂鼓岭梦园
SHIMAO GULING MENGYUAN

● **第五届银奖**
5th SILVER AWARD

建发·央玺
JIANFA · YANGXI

绿地香港拾野川
GREENLAND HONGKONG MOUNTAIN TIME

融创江南壹号院
SUNAC ONE PARK ERA

珠海恒荣·城市溪谷
ZHUHAI HENGRONG · CHENGSHI XIGU

深圳华侨城欢乐海岸
OCT BAY

业主单位：深圳华侨城都市娱乐投资公司
景观设计：SWA 集团
项目规模：1 250 000 m²
景观面积：xxx m²
建成时间：2012年
设计周期：61个月

Owner: OCT Urban Entertainment Investment Company of Shenzhen
Landscape Design: SWA Group
Project Scale: 1,250,000 m²
Landscape Area: xxx m²
Date of Completion: 2012
Cycle of Design: 61 months

Theme Park · 主题公园 ·

项目设计以教育、文化、娱乐、休闲为主导，通过创建一系列的游乐活动项目，鼓励公众参与其中。项目充分利用本地材料、绿色技术和可持续性手段，以提高节约资源的既定目标。设计理念以与水的互动为主线，水作为一种自然资源，美感元素、生态系统与交通运输辅助设施。该项目因其成功的平衡，成为中国综合项目的典范。

Locally sourced materials, green technologies and sustainable practices were utilized to enhance the resource responsible goals established for the project. The site concept is based on an interaction with water-as a natural resource, aesthetic element, ecological system, and conveyance amenity. The project has become a model for a successful approach to balanced mixed use project development in China.

Resort Hotel · 度假酒店 ·

黄山雨润涵月楼酒店采用传统徽南文化中建筑与景观的造型设计风格，创新性地将皖南村落的景观特征渗透到徽派建筑当中，以原生态的设计方式让酒店融于皖南山地。

Huangshan Yurun Hanyuelou Villan Resort adopts the architectural and landscape design style of the traditional Huinan culture, and innovatively penetrates the landscape features of the villages in southern Anhui into the Hui-style architecture and integrates the hotel into the mountains of southern Anhui with the original ecological design method.

黄山雨润涵月楼
HUANGSHAN YURUN HANYUELOU VILLA RESORT

业主单位：黄山松柏高尔夫乡村俱乐部有限公司
景观设计：贝尔高林国际（香港）有限公司
项目规模：165 000 m²
景观面积：165 000 m²
建成时间：2013年
设计周期：69个月

Owner: Huangshan Pine Golf & Country Club Co., Ltd.
Landscape Design: Belt Collins International (HK) Limited.
Project Scale: 165,000 m²
Landscape Area: 165,000 m²
Date of Completion: 2013
Cycle of Design: 69 months

上海万科翡翠滨江
SHANGHAI VANKE EMERALD RIVERSIDE

业主单位：万科集团
景观设计：CICADA；朗道
项目规模：60 000 m²
景观面积：45 969 m²
建成时间：2014年
设计周期：18个月

Owner: VANKE
Landscape Design: CICADA; LANDAU
Project Scale: 60,000 m²
Landscape Area: 45,969 m²
Date of Completion: 2014
Cycle of Design: 18 months

Residential Project · 居住项目 ·

万科翡翠滨江是目前上海内环内整个陆家嘴滨江沿线最大的城市综合体项目，在未来也将与老外滩、陆家嘴的顶级住宅区一同构成上海中央腹地的"黄金三角"。

The Vanke Emerald Riverside is currently the largest urban complex project along the Lujiazui Riverside within the inner ring of Shanghai. In the future, it will also form the "Golden Triangle" of Shanghai's central hinterland together with the top residential areas of the Old Bund and Lujiazui.

重庆东原 1891 印长江
CHONGQING DONGYUAN 1891

业主单位：重庆东原房地产有限公司
景观设计：SWA
项目规模：49 000 m²
景观面积：4 000 m²
建成时间：2015年
设计周期：12个月

Owner: Chong Qing Dong Yuan Real Estate Ltd.
Landscape Design: SWA
Project Scale: 49,000 m²
Landscape Area: 4,000 m²
Date of Completion: 2015
Cycle of Design: 12 months

第二届·金奖
2nd GOLD AWARD

Residential Project ·居住项目·

水景的连接创造了连续而变化的空间形式，以自然景色结合世外桃源故事中的桃花为灵感，引入长江与南山的自然景观，在城市中心创造一处世外桃源般不可替代的体验。

The waterscape connection creates a continuous but changing spatial form. Inspired by the natural scenery combined with the peach flowers mentioned in the legend of "retreat away from the world", the natural sceneries of the Yangtze River and Nan Mountain are introduced into this project, creating an irreplaceable experience for people who enjoy the city center like visiting a paradise.

第二届·金奖
2nd GOLD AWARD

Residential Project ·居住项目·

福州万科城
FUZHOU VANKE CITY

业主单位：福州市万科房地产有限公司
景观设计：SWA洛杉矶办公室
项目规模：450 000 m²
景观面积：350 000 m²
建成时间：2015年
设计周期：12个月

Owner: Fuzhou Vanke Real Estate Co., Ltd.
Landscape Design: SWA Los Angeles
Project Scale: 450,000 m²
Landscape Area: 350,000 m²
Date of Completion: 2015
Cycle of Design: 12 months

项目以"缝织"的景观意象为主要设计理念，除却整体规划层面组成交织的公共空间网络外，在实质的空间层面，步道、桥、廊、水景等区域也都以转折交缝的形式呈现出空间设计的独特性。

The main design concept of the project is to sew and weave the landscape. In addition to forming an intertwined public space network at the overall planning level, the physical space is laid out with the walkways, bridges, corridors, water, scenic spots in the form of transitions and joints, presenting the uniqueness of the space design.

万科云间传奇
LEGEND OF CLOUDS

业主单位：万科集团
景观设计：朗道国际设计
项目规模：57 432.5 m²
景观面积：42 464.3 m²
建成时间：2015年
设计周期：5个月

Owner: VANKE
Landscape Design: LANDAU
Project Scale: 57,432.5 m²
Landscape Area: 42,464.3 m²
Date of Completion: 2015
Cycle of Design: 5 months

第二届·金奖
2nd GOLD AWARD

Demonstration Area ·示范区·

本项目是万科集团2015年国际系精工别墅代表作之一，建筑风格和规划借鉴了松江本地里弄的元素，又和现代的新亚洲风格相结合，用简洁的、具有品质感的、人文的、实用的语言诠释居住的文化底蕴。

The design of Vanke Legend of Clouds, as one of the representative works of the International fine villa series of Vanke Group in 2015, absorbs the elements of the local lanes and alleys in Songjiang, combined with modern Asian style, illustrating the living culture in a simple, quality, humane and practical language.

莫奈大道 2.0
MONET AVENUE 2.0 AT VICTORIA GARDENS

业主单位：森林城市商业集团&西海岸商业发展公司
景观设计：SWA
项目规模：9 288 m²
景观面积：4 644 m²
建成时间：2015年
设计时间：8个月

Owner: Forest City Commerical Group
&West Coast Commercial Development
Landscape Design: SWA
design company
Project Scale: 9,288 m²
Landscape Area: 4,644 m²
Date of Completion: 2015
Cycle of Design: 8 months

Commercial Project · 商办项目 ·

莫奈大道2.0充分给予公众从电子商务中所无法获得的愉悦，以及令人难忘的购物体验。项目所配建的独特景观建筑就是为了创造出这样的体验，特别是在室外零售环境方面。

Monet Avenue 2.0 fully gives the public the unforgettable shopping experience that cannot be obtained from e-commerce. The unique landscape architecture of the project is to create a special outdoor retail environment.

北辰旭辉壹号院
CIFI CHATEAU

业主单位：旭辉集团 & 北辰地产
景观设计：广州山水比德设计股份有限公司
项目规模：18 000 m²
景观面积：16 000 m²
建成时间：2016年
设计周期：5个月

Owner: CIFI Group & Beijing North Star Co., Ltd.
Landscape Design: Guangzhou S.P.I Design Co., Ltd.
Project Scale: 18,000 m²
Landscape Area: 16,000 m²
Date of Completion: 2016
Cycle of Design: 5 months

Demonstration Area · 示范区 ·

本项目以苏式园林文化为底蕴和创作基调，提炼以"江南山秀"为表现的文化符号，以现代手法演绎江南传统园林意境，朝闻市井声，暮听林泉乐，让人陶醉在自然山水之间。

Based on the Suzhou garden culture, the project extracts the cultural symbol of "Jiangnan Mountains are elegant", illustrating the artistic conception of Jiangnan traditional gardens with modern techniques, letting its residents touch nature, and enjoy the life-style of hearing the sound of the city in the morning and listening to the melody sounds of trees and spring in the evening.

万科天空之城
UNI-CITY

业主单位：万科集团
景观设计：朗道国际设计
项目规模：47 916 m²
景观面积：5 106 m²（红线内）+13 360 m²（红线外）
建成时间：2016年
设计时间：3个月

Owner: VANKE
Landscape Design: LANDAU
Project Scale: 47,916 m²
Landscape Area: 5,106 m²(Inside the red line)+13,360 m²(Outside the red line)
Date of Completion: 2016
Cycle of Design: 3 months

Demonstration Area · 示范区 ·

万科天空之城是万科第一个在地铁上盖的商住办综合体首开区，场地规划和建筑设计为景观预留丰富的空间，满足销售需求的同时兼顾综合类产品的展示要求，弱化周边环境的不利影响。

As the first commercial-residential-office complex on the subway, the site planning and architectural design of UNI-CITY reserve ample space for the landscape to meet the demands for sales and the display of the comprehensive products, weakening the adverse effects of the surrounding environments.

远洋红岸澜山别墅庭院工程
SINO-OCEAN HONGANLANSHAN VILLA COURTYARD

业主单位：远洋集团
景观设计：小林治人
项目规模：86 666.66m²
景观面积：300 m²
建成时间：2016年
设计周期：0.5个月

Owner: Sino-Ocean Group Holding Limited.
Landscape Design: Haruto Kobayashi
Project Scale: 86,666.66 m²
Landscape Area: 300 m²
Date of Completion: 2016
Cycle of Design: 0.5 months

Residential Project · 居住项目 ·

这是一个纯日式风格的庭院改造项目，庭院四处景点分别巧妙地布置在几个空间内，客人在行走之间通过四个转折感受庭院的春花秋叶，夏日叠水溪流，坐观松枫秋叶，冬日观雪蓬莱，步移景异，院落美景，尽收眼底。

The project is a purely Japanese-style courtyard renovation. The four scenic spots are arranged in several spaces elaborately. When walking around, you can experience wonderful sceneries of four seasons through the four turning points. While walking through the courtyard, you will enjoy the summer creek, the autumn pine trees and maple leaves, and the winter snow. How amazing they are!

Commercial Project · 商办项目 ·

景观设计紧密结合建筑语汇，巧妙地将"黄浦江"引入项目当中，潺潺流动的水声和光影效果不仅激活了整个办公空间，同时也为该项目打造了独一无二的标志性景观特征。

Combined with the architectural language, the landscape design skillfully brings the Huangpu River into this project. The running water and the light effects not only activate the entire office space, but also create a unique and iconic landscape feature for the project.

BUND SOHO 外滩 SOHO

业主单位：SOHO中国
景观设计：澳洲艺普得城市设计咨询有限公司
项目规模：189 508 m²
景观面积：12 000 m²
建成时间：2017年
设计周期：36个月

Owner: SOHO China
Landscape Design: INTEGRATED PLANNING AND DESIGN PTY LTD
Project Scale: 189,508 m²
Landscape Area: 12,000 m²
Date of Completion: 2017
Cycle of Design: 36 months

重庆龙湖舜山府展示区
CHONGQING LONGFOR SHUNSHAN MANSION SALES CENTER

业主单位：重庆龙湖
景观设计：SWA旧金山
项目规模：284 625 m²
景观面积：35 000 m²（展示区）
　　　　　63 000 m²（大区一期）
建成时间：2017年
设计周期：7个月

Owner: Longfor Chongqing
Landscape Design: SWA San Francisco
Project Scale: 284,625 m²
Landscape Area: 35,000 m² (Demonstration Area)
　　　　　　　　 63,000 m² (Phase I)
Date of Completion: 2017
Cycle of Design: 7 months

Demonstration Area · 示范区 ·

本项目借用"山门"的概念，强调"进入"的仪式感，设置若干转折观景点令人饱览山谷风貌，最大化人与山的互动。展示区依托天然断崖呈现"山林之镜"的倒影水景，将天色刻录在水之镜中，美景尽收眼底。

Adopting the concept of "Mountain Gate", emphasizing on the ceremonial atmosphere of "entering", the project takes advantage of the natural cliff setting to create several turning points. All the beautiful changes between sunny sky, rainy days, misty forests, and cloudy hills will be engraved in this mirror of water and revealed in front of visitors.

福州万科仓前九里项目
FUZHOU VANKE CANGQIANJIULI PROJECT

业主单位：福州市万滨房地产有限公司
景观设计：Lab D+H
项目规模：8 996 m²
景观面积：8 996 m²
建成时间：2017年
设计周期：5个月

Owner: Fuzhou Wanbin Real Estate Co., Ltd.
Landscape Design: Lab D+H
Project Scale: 8 996 m²
Landscape Area: 8 996 m²
Date of Completion: 2017
Cycle of Design: 5 months

第四届·金奖
4th GOLD AWARD

Urban Regeneration ·城市再生·

这是一个完全融入当地历史文化的景观项目，用一种更加自然的方式复兴烟台山，在修复和翻新旧貌的同时，把设计本身藏起来，只让游客沉浸在整个场所营造之中。

To let the landscape design project fully integrate into the local history and culture, the revival of Yantai Mountain should use a more natural, innate design method to repair the site and integrate design into the environment.Inaddition，to make people immersed in the atmosphere created by this place, the design itself is invisible.

第四届·金奖
4th GOLD AWARD

Residential Project ·居住项目·

龙湖新江与城清晖岸
LONGFOR NEW RIVER AND CITY QINGHUI SHORE

业主单位：重庆龙湖怡置地产发展有限公司
景观设计：WTD纬图设计
项目规模：235 300 m²
景观面积：62 700 m²
建成时间：2017年
设计时间：6个月

Owner: Chongqing Longfor Yi Zhen Real Estate Development Co., Ltd.
Landscape Design: WTD
Project Scale: 235,300 m²
Landscape Area: 62,700 m²
Date of Completion: 2017
Cycle of Design: 6 months

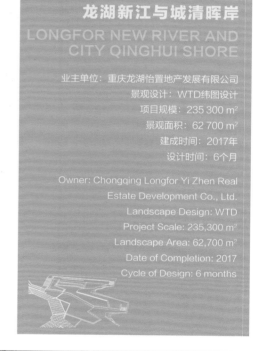

本项目以一条贯穿全园的运动跑道为主线，串联起其他景观结构，实现环道对全园功能的连接，营造充满青春时尚格调和幸福浪漫氛围的人居环境。

Taking a sports track running through the whole park as the main line, connecting other landscape structures in series, the project realizes the connection of the ring road to the functions of the whole park, and creates a living environment full of youthful fashion style and happy romantic atmosphere.

龙湖旭辉春江悦茗
LONGFOR CHUNJIANG CENTRAL

业主单位：杭州龙卓房地产开发有限公司
景观设计：安道设计
项目规模：143 955.8 m²
景观面积：30 409 m²
建成时间：2017年
设计周期：24个月

Owner:
Hangzhou Longzhuo Real Estate Development Co., Ltd.
Landscape Design: antao
Project Scale: 143,955.8 m²
Landscape Area: 30,409 m²
Date of Completion: 2017
Cycle of Design: 24 months

第四届·金奖
4th GOLD AWARD

Residential Project ·居住项目·

本项目中温润优雅的法式人文社区用极简完美的比例完成"建筑肌理"与"景观肌理"的对话，为每一个庭院镌刻专属花语，所有的精致只为让客户感受细节与浪漫、奢华与品质。

The project is a warm and elegant human community featured with French luxury. The designers use simple and perfect proportion to complete the dialogue between "architectural texture" and "landscape texture". The designers design for every courtyard in its' exclusive flower language, special household pattern and identified sculpture. The designers provide guests with the details and romance, the luxury and the quality.

Culture Project ·文化项目·

泸州·顺成和庄园
LUZHOU SHUNCHENGHE MANOR

业主单位：泸州顺成和投资有限公司
景观设计：重庆尚源建筑景观设计有限公司
景观面积：12 337 m²
建成时间：2016年
设计周期：12个月

Owner: Luzhou Shunchenghe Investment Co., Ltd.
Landscape Design:
CHONGQING ADVO – NATURE ARCHITECTURE
& LANDSCAPE DESIGN
Landscape Area: 12,337 m²
Date of Completion: 2016
Cycle of Design: 12 months

项目地理位置优越，崖顶视角可远眺长江。设计师依据建筑线条的多层次，结合当地闪电雷区的地域属性，利用被破坏的斜坡塑造闪电造型。一条折线是消弥高差的步行道，另一条折线是雨水收集、净化曝氧的水道。

The location of Luzhou Shunchenghe Manor is superior and sits on the cliff top. It has a very good perspective that can overlook the Yangtze River. The project is based on the characteristics of the architecture form, dynamic and rich in layers, combined with the geographical attributes of the local lightning thunder area. One folding line is the walkway that digests the height difference, and the other is the waterway that collects and purifies the rain water.

音昱水中天
SANGHA BY OCTAVE

业主单位：苏州万邦置业发展有限公司
景观设计：地茂景观设计咨询（上海）有限公司
景观面积：40 826 m²
建成时间：2017年
设计周期：34个月

Owner:
Suzhou Wanbang Real Estate Development Co., Ltd.
Landscape Design: Design Land Collaborative Ltd.
Landscape Area: 40,826 m²
Date of Completion: 2017
Cycle of Design: 34 months

Cultural Tourism Project ·文旅项目·

音昱水中天是一个真正集生活、工作、学习为一体的综合型社区，强大的社区体系致力于为住户提供特色课程设计及一系列医疗保健服务。

Sangha by Octave is a comprehensive community that truly integrates life, work, and study. A strong community system is dedicated to providing residents with unique curriculum design and a series of medical care services.

Residential Project ·居住项目·

贵阳旭辉·观云
SKYLINE

业主单位：贵阳辉沛企业管理有限公司
景观设计：朗道国际设计
项目规模：252 817.51 m²
景观面积：50 654 m²
建成时间：2018年
设计周期：3个月

Owner: CIFI Group
Landscape Design: LANDAU
Project Scale: 252,817.51 m²
Landscape Area: 50,654 m²
Date of Completion: 2018
Cycle of Design: 3 months

本项目用景观连接现代生活与生态园景，运用贵阳竹文化元素来营造景观序列，将贵州的远山与浮云化为园景脉络，引光与影入景，构建生态自然与现代工艺相结合的美好空间场所。

The design uses landscapes to connect modern life and ecological landscapes and uses Guiyang bamboo culture elements to create landscape sequences, transforming Guizhou's distant mountains and floating clouds into landscapes. The whole model constructs a beautiful space combined with ecological nature and modern technology.

杭州萧山 前湾 国际社区·无界公园
XIAOSHAN HANGZHOU BAY UNBOUNDED PARK

业主单位：仁恒置地&香港置地
景观设计：上海广亩景观设计有限公司
景观面积：27 961 m²
建成时间：2018年
设计周期：9个月

Owner: Yanlord Land&HongKong Land
Landscape Design:
Shanghai GM Landscape Design Co., Ltd.
Landscape Area: 27,961 m²
Date of Completion: 2018
Cycle of Design: 9 months

Demonstration Area ·示范区·

本项目以"钱江潮"文化元素为设计出发点，通过精巧的空间布局形式、格调高雅的艺术品、富有质感的材料提升整体景观格调，打造开放包容、平等友好的无界公园。

Taking the "Qianjiang Tide" cultural elements as the design starting point, the overall landscape style is enhanced through the exquisite spatial layout, elegant artworks, and textured materials to create an open, inclusive, equal and friendly unbounded park.

Urban Regeneration ·城市再生·

宁波万科芝士公园
VANKE CHEESE PARK IN NINGBO

业主单位：宁波万科企业有限公司
景观设计：WTD纬图设计
项目规模：35 000 m²
景观面积：27 000 m²
建成时间：2018年
设计时间：6个月

Owner: Ningbo Vanke Co., Ltd.
Landscape Design: WTD
Project Scale: 35,000 m²
Landscape Area: 27,000 m²
Date of Completion: 2018
Cycle of Design: 6 months

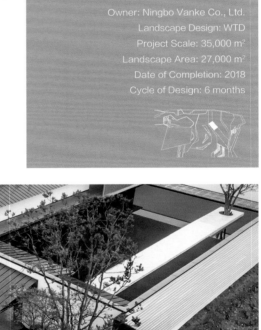

项目力求打造宁波最具影响力的素质教育综合体，并成为宁波的城市地标之一。设计以"重力场"为主题，激发孩子们对科学知识的好奇心，创建标志性的"芝士猫"形象，树立鲜明的场所身份认知。

The project strives to build the most influential quality education complex, and make it one of the landmarks in Ningbo. Based on the theme of "Gravity Field", which stimulates children to be curious about scientific knowledge, the design builds the iconic image of "Cheese Kitty", creating a distinctive identity.

万科·桂语东方
LAUREL ORIENTAL · VANKE

业主单位：万科地产&远洋地产&卓越地产
景观设计：安道设计
景观面积：12 800 m²
建成时间：2018年
设计时间：2个月

Owner: Vanke&Sino-Ocean&Excellence
Landscape Design: antao
Landscape Area: 12,800 m²
Date of Completion: 2018
Cycle of Design: 2 months

Demonstration Area ·示范区·

项目用高度抽象的现代主义充分诠释了江南烟雨生活的多样遐想，打破"中国风"惯用的曲径通幽，极致收敛的设计手法将白墙、青瓦、黑砂、翠绿等元素共同融入这座庭院的全部空间。

The Laurel Oriental uses the highly abstract minimalist modernism, which can fully show diverse reveries about Jiangnan life. It breaks the conventional Chinese style, which has the winding streets and the deep courtyards. The four elements covering white walls, grey tiles, black sand, and bright greenness, make up the whole space of this yard.

Office Building ·办公楼·

中洲控股金融中心
SCC

业主单位：深圳市中洲投资控股股份有限公司
景观设计：SWA索萨利图
项目规模：232 690.41 m²
景观面积：24 850 m²
建成时间：2016年
设计时间：120个月

Owner: Shenzhen Centralcon
Landscape Design: SWA Sausalito
Project Scale: 232,690.41 m²
Landscape Area: 24,850 m²
Date of Completion: 2016
Cycle of Design: 120 months

第四届·银奖
4th SILVER AWARD

本项目将植物带来的热带感受延伸至连廊里的垂直绿化，提高平面绿意到建筑的层次；高度整合场地元素，追求价值的最大化，多样景观元素的利用与协调彰显整体设计的融合和高端，兼具优雅与实用性。

The tropical feeling brought by the plants is extended to the vertical greening in the corridor, which lifts the flat greening to the building. Combined with site elements, this project, pursuing maximum value, skillfully uses the diverse landscape elements to demonstrate the integrated, high-end, elegant, practical, overall design.

北京融创·壹号院
SUNAC ONE SINO PARK IN BEIJING

业主单位：融创中国控股有限公司北京公司
景观设计：贝尔高林国际（香港）有限公司
项目规模：28 200 m²
景观面积：16 300 m²
建成时间：2017年
设计时间：36个月

Owner: Sunac Group
Landscape Design:
Belt Collins International (Hong Kong) Limited.
Project Scale: 28,200 m²
Landscape Area: 16,300 m²
Date of Completion: 2017
Cycle of Design: 36 months

第四届·银奖
4th SILVER AWARD

Residential Project ·居住项目·

本项目从枯燥的线性布局中跳出来，创造精彩趣味的空间，平面由最初的中轴线开放空间大草坪演变为错落有致的趣味像素花园，力求为居者营造一份具有现代特色花园的住宅感受。

To avoid the boring linear layout and create wonderful and interesting spaces, the designers made many different attempts. The plan has evolved from the original large lawn with open space on the central axis to a well-spaced and interesting pixel garden, striving to create a feeling of living a modern garden for its residents.

第四届·银奖
4th SILVER AWARD

Demonstration Area ·示范区·

承德金山岭洛嘉山精灵松塔乐园
CHENGDE JINSHANLING V-ONDERLAND WITH MONTAIN ELF AND PINE CONE

业主单位：滦平坤元房地产开发有限公司
景观设计：奥雅设计
项目规模：2 300 m²
景观面积：2 300 m²
建成时间：2017年
设计时间：6个月

Owner: Luanping Kunyuan Real Estate Development Co., Ltd.
Landscape Design: L&A Design
Project Scale: 2,300 m²
Landscape Area: 2,300 m²
Date of Completion: 2017
Cycle of Design: 6 months

项目场地位于山体坡地，周边群山环绕。受满地松果启发，设计团队将创造一个充分尊重地形的自然乐园、来自大自然的童话世界，调动孩子们各系感官自发探索，打造极富脑洞的冒险胜地。

The site is located in a mountain slope, surrounded by mountains. Inspired by the pinecones, the design team has designated the theme park as a natural paradise full of respect for mountainous terrain. The designers want to design a growth destination where children can experience nature and play freely.

世茂鼓岭梦园
SHIMAO GULING MENGYUAN

业主单位：世茂集团
景观设计：上海广亩景观设计有限公司
项目规模：79 567.2 m²
景观面积：106 489.4 m²
建成时间：2017年
设计周期：2个月

Owner: ShiMao Group
Landscape Design:
Shanghai GM landscape design Co., Ltd.
Project Scale: 79,567.2 m²
Landscape Area: 106,489.4 m²
Date of Completion: 2017
Cycle of Design: 2 months

第四届·银奖
4th SILVER AWARD

Residential Project ·居住项目·

项目秉承尊重自然，与环境协调统一的设计理念，最大限度地保护了场地原生植物与土地肌理，结合闽南当地历史文脉，以当代的手法再现富有诗意的中国院落式山居住宅，营造人与环境共生的绿色健康生活。

This project adhered to the design concept of respecting nature and environment, protected the original plants and soil texture of the site to solve out the problem of stormwater management. The project takes the contemporary design technique to represent the poetic Chinese courtyard style mountain house and creates a green and healthy life of symbiosis between human-being and environment.

建发·央玺
JIANFA · YANGXI

业主单位：建发房地产集团
景观设计：成都赛肯思创享生活景观设计股份有限公司
项目规模：125 481 m²
景观面积：17 013 m²
建成时间：2017年
设计时间：24个月

Owner: JianFa Real Estate Group
Landscape Design: Chengdu SecondNature INNOJOY Landscape Design Co., Ltd.
Project Scale: 125,481 m²
Landscape Area: 17,013 m²
Date of Completion: 2017
Cycle of Design: 24 months

第五届·银奖
5th SILVER AWARD

Residential Project ·居住项目·

福建好古，建发央玺凭古为趣。本项目从"桃花源"的意境出发，将"居山水之间"的传统处居思想与"出则繁华，入则幽静"的现代人居追求完美糅合，探索古典园林在当今时代的全新演绎。

Because Fujianese are favor of the traditional style, this project exactly presents the ancient traits to meet their desires. Inspired by Taohuayuan which refers to the name of the place originated from the ancient poem, this project combines the traditional ideal living place with the modern one, exploring the new interpretation of classical gardens in the contemporary era.

第五届·银奖
5th SILVER AWARD

Demonstration Area ·示范区·

绿地香港拾野川
GREENLAND HONGKONG MOUNTAIN TIME

业主单位：绿地香港控股有限公司
景观设计：Lab D+H
项目规模：800 000 m²
景观面积：32 000 m²
建成时间：2018年
设计时间：2.5个月

Owner: Greenland HK
Landscape Design: Lab.D+H
Project Scale: 800,000 m²
Landscape Area: 32,000 m²
Date of Completion: 2018
Cycle of Design: 2.5 months

项目设计源于场地的独特条件，通过路径化的空间布置以及叙事感的氛围引导，讲述一个返璞归真的故事——归源记。一个故事，四行小诗，八个源景，无数回忆。

The design originated from the unique conditions of the site, through the path of space layout, and the atmosphere of narrative guidance, Shiyechuan is looking forward to telling a story of returning to the truth-return Source note, covering a story, four lines of poetry, eight source scenes, and countless memories.

Demonstration Area · 示范区 ·

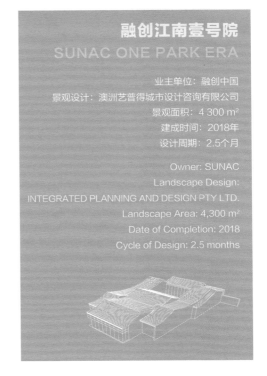

融创江南壹号院
SUNAC ONE PARK ERA

业主单位：融创中国
景观设计：澳洲艺普得城市设计咨询有限公司
景观面积：4 300 m²
建成时间：2018年
设计周期：2.5个月

Owner: SUNAC
Landscape Design:
INTEGRATED PLANNING AND DESIGN PTY LTD.
Landscape Area: 4,300 m²
Date of Completion: 2018
Cycle of Design: 2.5 months

"人"字形斜坡屋顶与极简的白墙、黛瓦元素给人以鲜明的视觉印象，形成连续而富有韵律的整体界面。建筑与景观的结合刻画了一幅江南山水的绝妙画卷。

The herringbone sloping roofs, and the minimalist white walls and black tiles elements give a vivid visual impression, forming an overall continuous and rhythmic picture. The building and the landscape together create a wonderful picture of mountains and rivers in Jiangnan.

珠海 恒荣·城市溪谷
ZHUHAI HENGRONG · CHENGSHI XIGU

业主单位：深圳市恒荣置地有限公司
景观设计：贝尔高林国际（香港）有限公司
景观面积：50 000 m²
建成时间：2018年
设计周期：11个月

Owner: Shenzhen Heng Rong Co., Ltd.
Landscape Design: Belt Collins International
(Hong Kong) Co., Ltd.
Landscape Area: 50,000 m²
Date of Completion: 2018
Cycle of Design: 11 months

Residential Project · 居住项目 ·

项目以天桥和连廊作为依托，充分利用空中连廊带来的空间趣性，将小区人车动线分割，上有连廊，下有车道，将城市溪谷雕琢成了一处完美的4D盛景。

The project relies on sky bridges and corridors by making full use of the spatial interest brought by the corridors in the sky. It separates the traffic from people in the community, with a corridor on the top and a driveway below, turning the urban valley into a perfect 4D grand scene.

04 室内设计 INTERIOR DESIGN

第一届金奖

1st GOLD AWARD

香港希慎广场
HYSAN PLACE, HONG KONG

第二届金奖

2nd GOLD AWARD

济南阳光一百艺术馆
JINAN SUNSHINE 100 MUSEUM OF ART

虹桥天地
THE HUB

第三届金奖

3rd GOLD AWARD

格柏购物中心
GERBER SHOPPING MALL

杭州万科随园嘉树护理院
SUIYUAN HURSING HOME

京都四季酒店全日餐厅
BRASSERIE RESTAURANT FOUR SEASONS KYOTO

金沙不纸书店
GOLDEN SAND NO PAPER BOOKSTORE

万科·美好家·无限系—115户型精装产品设计研发
VANKE GOOD HOME INFINITE SERIES PRODUCT LINE R&D—115 TYPE BOUTIQUE ROOM

第四届金奖

4th GOLD AWARD

雪月花日本料理
JAPANESE WAKA, HAIKU & SETSUGEKKA

杭州万科良渚文化村郡西云台售楼处
HANGZHOU VANKE JUNXI MOUNTAIN VILLA SALES CENTRE

华侨城苏河湾上海宝格丽酒店
OCT SUHE CREEK BVLGARI HOTEL

华发广钢新城项目
HUAFA GUANGGANG NEW TOWN

第五届金奖
5th GOLD AWARD

文舺
MONKA HOTEL

西打磨厂共享际
GRINDING FACTORY 5LMEET

上海滨江道办公楼
IN-BUND OFFICE, SHANGHAI

第四届银奖
4th SILVER AWARD

所见西溪度假酒店
SAVOIR RESORT

有朋青年共享社区（畅景公寓改造）
UPON COMMUNITY(CHANGJING APARTMENT RENOVATION DESIGN)

郑州阳光城·檀悦售楼处
WAVING RIBBON

中鹰·黑森林"多彩空间"样板房
ZHONGYING BLACK FOREST 'CHROMATIC SPACES' SHOW APARTMENT

第五届银奖
5th SILVER AWARD

广州金地壹阅府
GUANGZHOU GEMDALE THE ONE VILLA

十字轴线之间
CHARMING EASTERN

蔚来中心 / 杭州西湖
NIO HOUSE / HANGZHOU WEST LAKE

伍兹贝格上海工作室
WOODS BAGOT SHANGHAI STUDIO

行云琉光
LIGHTING IN THE WAVES

香港希慎广场
HYSAN PLACE, HONG KONG

业主单位：希慎兴业
室内设计：Benoy贝诺
项目规模：110 000 m²
建成时间：2013年

Owner: Hysan Development Company Limited.
Interior Design: Benoy
Project Scale: 110,000 m²
Date of Completion: 2013

Shopping Mall ·购物中心·

第一届·金奖
1st GOLD AWARD

希慎广场作为大中华区域首个获得LEED最高白金级别认证的购物中心发展项目，是当地最具特色的综合购物中心之一，现今这座高17层的购物中心已成为铜锣湾商业区的地标性建筑。

Hysan Place, as the first shopping mall development project in the Greater China region to obtain the highest LEED platinum level certification, is one of the most distinctive integrated shopping malls in the local area. Today this 17-storey shopping mall has become a landmark building in the Causeway Bay business district.

济南阳光一百艺术馆
JINAN SUNSHINE 100 MUSEUM OF ART

业主单位：济南阳光一百房地产开发有限公司
室内设计：深圳市派尚环境艺术设计有限公司
室内面积：2 888 m²
建成时间：2013年

Owner: Jinan Sunshine 100 Real Estate Development Co., Ltd.
Interior Design: Shenzhen Panshine Interior Design Co., Ltd.
Landscape Area: 2,888 m²
Date of Completion: 2013

第二届·金奖
2nd GOLD AWARD

Sales Center ·售楼中心·

本项目以"艺术会馆里的售楼处"为规划目标，构筑艺术气息和品质感氛围，深化设计赋予空间独特气质的途径。多层次的设计勾勒出具有东方禅意的空间轮廓，多组原创的艺术装置强化出会馆的主题性。

The project aims to become "a sales department in an art hall" in its space plan, build an atmosphere full of art flavor and sense of quality. The multi-level design creates a noble and highly striking atmosphere while multiple groups of innovative art devices are used to strengthen the themes of the hall.

第二届·金奖
2nd GOLD AWARD

虹桥天地
THE HUB

业主单位：瑞安房地产&中国新天地
室内设计：CallisonRTKL
项目规模：170 000 m²
建成时间：2015年
设计周期：24个月

Owner: Shui On Land&China XTD
Interior Design: CallisonRTKL
Project Scale: 170,000 m²
Date of Completion: 2015
Cycle of Design: 24 months

·TOD·

项目以"连接"为设计主旨，建立人与人之间从物理到情感的维度连接，将绿色环保延续到室内设计当中，结合天顶采光、木纹饰面、室内绿植、流线造型的设计元素，打造城市绿洲的休闲氛围。

The design theme of the project is "connection", establishing the connection among people from the physics to emotion. In addition, the concept of green environments has been taken into this project, and combined with the other design elements, such as dome ceiling, woodgrain finishes, greenery decorations, and streamlining, creating the natural and comfortable experience for consumers.

Shopping Mall · 购物中心 ·

格柏购物中心
GERBER SHOPPING MALL

业主单位：凤凰房地产开发有限公司
室内设计：IFG 伊波莱茨建筑设计
项目规模：120 000 m²
室内面积：5 500 m²
建成时间：2014年
设计周期：44个月

Owner:
Phoenix Real Estate Development GmbH
Interior Design:
Ippolito Fleitz Group-Identity Architects
Project Scale: 120,000 m²
Interior Area: 5,500 m²
Date of Completion: 2014
Cycle of Design: 44 months

项目通过高档次的内部装修吸引各个层次的顾客群体，成为斯图加特的地标性建筑。高级的环境质量、清晰的导向系统和独特的装饰元素都将成为这座中心的代表特征。

The project attracts customers at all levels through high-end interior decoration, and becomes a landmark building in Stuttgart. High-level environmental quality, clear guidance system and unique decorative elements will all become the representative features of this center.

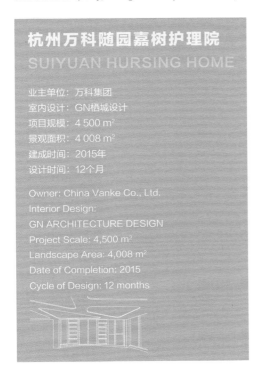

杭州万科随园嘉树护理院
SUIYUAN NURSING HOME

业主单位：万科集团
室内设计：GN栖城设计
项目规模：4 500 m²
景观面积：4 008 m²
建成时间：2015年
设计时间：12个月

Owner: China Vanke Co., Ltd.
Interior Design:
GN ARCHITECTURE DESIGN
Project Scale: 4,500 m²
Landscape Area: 4,008 m²
Date of Completion: 2015
Cycle of Design: 12 months

Healthcare Project · 医养项目 ·

项目以长者行为特点和人体工学研究为基础，进行针对性、系统化的室内设计，为老人们提供集多样化健康管理、生活服务空间等于一体的专业型高品质长者护理机构，打造万科集团第一个护理型养老机构。

Based on the elderly behavior and ergonomics, this project carries out the targeted and systematic interior design, providing the elderly with a professional and high-quality nursing house where the elderly can enjoy diversified health management and life services, building the first nursing care home owned by Vanke Group for the elderly.

Restaurant · 餐饮 ·

京都四季酒店全日餐厅
BRASSERIE RESTAURANT FOUR SEASONS KYOTO

业主单位：京都东山酒店资产TMK
室内设计：Kokaistudios
项目规模：870 m²
室内面积：870 m²
建成时间：2016年
设计周期：24个月

Owner:
Kyoto Higashiyama Hospitality Assets TMK
Interior Design: Kokaistudios
Project Scale: 870 m²
Interior Area: 870 m²
Date of Completion: 2016
Cycle of Design: 24 months

京都四季酒店景色优美，依山傍水，整体项目建造采用鲜明的建筑策略，汲取京都传统的室内外连接系统，利用挑高的框景手法展示户外景观。

Brasserie Restaurant Four Seasons Kyoto has beautiful scenery and is surrounded by mountains and rivers. The overall project construction adopts a distinctive architectural strategy, absorbing the traditional indoor and outdoor connection system of Kyoto, and using the high-height frame method to display the outdoor landscape.

金沙不纸书店
GOLDEN SAND NO PAPER BOOKSTORE

业主单位：成都太行瑞宏房地产开发有限公司　　Owner: Chengdu Taihang Rui Hong Real Estate Development Company Limited
室内设计：XY+Z DESIGN　　Interior Area: XY+Z DESIGN
项目规模：5 000 m²　　Project Scale: 5,000 m²
室内面积：2 383 m²　　Interior Area: 2,383 m²
建成时间：2016年　　Date of Completion: 2016
设计周期：5个月　　Cycle of Design: 5 months

Sales Center · 售楼中心 ·

第三届·金奖
3rd GOLD AWARD

设计将"纸"元素融入其中，运用不同元素实现"售楼+办公+书店"的多重空间体验，结合智能照明系统、垂直绿化和声环境控制等，打造集绿色、健康、科技、人文、时尚为一体的文化聚集地。

The design emphasized the element "paper" and created a multi-experience space: "sales+office+bookstore" with different kinds of elements. The project is a combination of intelligent lighting system, vertical greening, sound environment control, and cultural display, creating a cultural gathering place featured with green, health, technology, humanities and fashion.

万科·美好家·无限系—115户型精装产品设计研发
VANKE GOOD HOME INFINITE SERIES PRODUCT LINE R&D—115 TYPE BOUTIQUE ROOM

业主单位：天津万科
室内设计：CLV赛拉维
项目规模：115 m²
室内面积：115 m²
建成时间：2016年
设计周期：12个月

Owner: Tianjin Vanke
Interior Design: CLV.DESIGN.
Project Scale: 115 m²
Landscape Area: 115 m²
Date of Completion: 2016
Cycle of Design: 12 months

Model Room · 样板房 ·

第三届·金奖
3rd GOLD AWARD

整个户型以一根结构柱作为建筑核心，通过楼板加固加厚，与建筑外剪力墙结构结合，满足其力学要求。设计师通过对不同客群的需求进行研究，按照不同家庭结构、生活习惯、生活人数，将产品拆分。

This structural column is the core of the building, reinforced and thickened by the floor slab, and combined with the outer shear wall structure of the building to meet its mechanical requirements. To satisfy the needs of different customer groups, the products are split based on different family structures, living habits, and the number of living people.

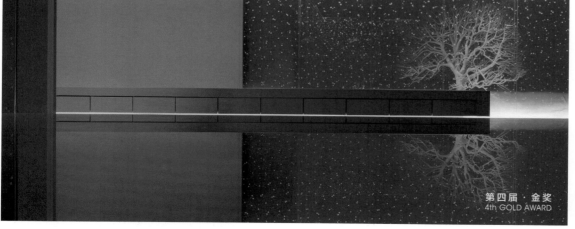

第四届·金奖
4th GOLD AWARD

Restaurant · 餐饮 ·

雪月花日本料理
JAPANESE WAKA, HAIKU & SETSUGEKKA

业主单位：长春市原味日本料理
室内设计：上海黑泡泡建筑装饰设计工程有限公司
项目规模：1 300 m²
室内面积：1 300 m²
建成时间：2016年
设计周期：4个月

Owner:
Changchun Ajimi Original Japanese Cuisine
Interior Design:
Shanghai Hipp Architectural Decoration Design Co., Ltd.
Project Scale: 1,300 m²
Interior Area: 1,300 m²
Date of Completion: 2016
Cycle of Design: 4 months

项目主设计师曾语：我们无法拒绝来自生存建筑、居住室内所带给你的潜在影响，相比之下似乎"传达什么"比争论"这是什么"更有意义，这也是设计师将这间料理店的名字改为"雪月花"的缘起。

As the chief designer of the project said: We cannot refuse the potential impact from living buildings and living interiors, so it seems that "what is this building conveyed" is more meaningful than arguing about "what is this building". That's why the designer gave the name of the restaurant "Setsugekka".

杭州万科良渚文化村郡西云台售楼处
HANGZHOU VANKE JUNXI MOUNTAIN VILLA SALES CENTRE

业主单位：杭州万科	Owner: Hangzhou Vanke
室内设计：木君建筑设计	Interior Design: MDO
项目规模：75 768.24 m²	Project Scale: 75,768.24 m²
室内面积：400 m²	Interior Area: 400 m²
建成时间：2017年	Date of Completion: 2017
设计时间：2个月	Cycle of Design: 2 months

Sales Center · 售楼中心 ·

项目意在优美的自然环境中创造光影、比例和体验相平衡的永恒空间。中央被分为模型区和长吧台，通过折叠的天花巧妙连接，平缓地流向室外水景。从画作到设计，项目动态地呈现传统屋顶元素。

The project intends to create a timeless space balanced between light and shadow, proportion and experience in a beautiful natural environment. The center is divided into a model area and a long bar, which are skillfully connected by a folding ceiling, gently extending to the outdoor waterscape. From painting to design, the traditional roof elements are dynamically presented.

华侨城苏河湾上海宝格丽酒店
OCT SUHE CREEK BVLGARI HOTEL

业主单位：华侨城（上海）置地有限公司
室内设计：
Antonio Citterio Patricia Viel Interiors S.r.l
BHD Consulting Limited
项目规模：234 091 m²
室内面积：23 800 m²
建成时间：2017年
设计周期：60个月

Owner: OCT Land (Shanghai) Investment Ltd.
Interior Design:
Antonio Citterio Patricia Viel BHD Consulting Limited
Project Scale: 234,091 m²
Interior Area: 23,800 m²
Date of Completion: 2017
Cycle of Design: 60 months

Hotel · 酒店 ·

项目是上海城市中心新地标，属于世界级滨水城市复兴。项目以现代的手法将迎宾大道水纹铺装肌理延续到整个广场，在镜面水景中倒映历史的文化根系，让客人充分感受到文化底蕴的传承。

The project is a new landmark in the central city of Shanghai, revitalizing the world-class waterfront city. The water pattern paving texture of Yingbin Avenue is extended to the entire square with modern techniques, reflecting the history in the mirrored waterscape, allowing guests to fully feel the cultural heritage.

华发广钢新城项目
HUAFA GUANGGANG NEW TOWN

业主单位：珠海华发实业股份有限公司	Owner: Zhuhai Huafa Industrial Share Co., Ltd.
室内设计：威尔逊室内建筑设计公司	Interior Design: Wilson Associates
项目规模：1 891 m²	Project Scale: 1,890 m²
室内面积：400 m²	Interior Area: 400 m²
建成时间：2017年	Date of Completion: 2017
设计周期：12个月	Cycle of Design: 12 months

Residential Project · 居住项目 ·

设计师采用独立家品增添空间感，使现代感和经典墙壁设计相平衡，打造"现代版的经典"，使华发广钢新城项目展示出瑰丽堂皇的气派。室内设计特别以黑白色为主调，展示现代崭新的家品。

The designer uses independent furniture to add a sense of space to this project, balancing the design of the modern wall and the classic one to create a "modern version of the classic", and making this project grand. The key tone of the interior design is black and white, presenting a brand-new modern furniture.

第五届·金奖
5th GOLD AWARD

Homestay ·民宿·

文舺
MONKA HOTEL

业主单位：文舺行旅
室内设计：境观室内装修设计有限公司
项目规模：290.8161 m²
室内面积：242.5878 m²
建成时间：2017年
设计周期：3个月

Owner: Monka Hotel
Interior Design: HAD Interior Design Co., Ltd.
Project Scale: 290.8161 m²
Interior Area: 242.5878 m²
Date of Completion: 2017
Cycle of Design: 3 months

项目用归零的态度重新认识文舺的地理基因和文化脉络，不着痕迹地模糊传统和现代的界线，灰色的清水混凝土加上大小不一的黑色窗格，以不彰显自己的姿态安静地凝视着身旁的老建筑。

To start the case, the project reset the geographic and cultural contest of the location of the case to default. The designers attempted to put ambiguity on the border between traditional architecture and modern styles without leaving any traces. With grey architectural concrete plus black windowpane of different sizes, this building quietly stares at the adjacent old buildings without highlight but with the attitude full of humbleness.

西打磨厂共享际
GRINDING FACTORY 5LMEET

业主单位：优享创智（北京）科技服务有限公司
室内设计：大观建筑设计DAGA Achitects
室内面积：520 m²
建成时间：2017年
设计周期：1个月

Owner: 5Lmeet
Interior Design: DAGA Achitects
Landscape Area: 520 m²
Date of Completion: 2017
Cycle of Design: 1 month

第五届·金奖
5th GOLD AWARD

Urban Regeneration ·城市再生·

北京西打磨厂胡同内的这处小院子原本是"民国"年间的瑞华染料行，经过改造设计，这处具有历史意义的传统小院儿已然成功转变为北京城内一处极具吸引力的长租公寓。

The small courtyard in GRINDING FACTORY 5LMEET Hutong in Beijing was originally a Ruihua Dye Shop in the Republic of China. After renovation and design, this historically significant traditional courtyard has been successfully transformed into an attractive long-term rental apartment in Beijing.

第五届·金奖
5th GOLD AWARD

Office Building ·办公项目·

上海滨江道办公楼
IN-BUND OFFICE, SHANGHAI

业主单位：印力集团联合德普置地
室内设计：HPP建筑事务所
室内面积：2 875 m²
建成时间：2018年
设计周期：6个月

Owner: SCPG & Development.Co., Ltd.
Interior Area: HPP Architects
Interior Area: 2,875 m²
Date of Completion: 2018
Cycle of Design: 6 months

这座历经百年历史的保护建筑位于上海黄浦江东岸，为保留建筑原有的风貌，并满足现代商业与办公采光需求，HPP设计师将立面和室内空间在原有基础上重新划分。以深红色砖块为元素，结合多种拼砌工艺，强调出新老建筑的延续和呼应，保留了整个空间的工业感。

The In-Bund Office is located on the east bank of the Huangpu River with a century-long history. The architects have re-designed the facade to let natural light flood in, so as to enhance the interior condition as a modern business office while retaining its original style. The building has deep red bricks as its major design elements that were laid with a pattern, creating a unique style manifesting the combination of old and new.

所见西溪度假酒店
SAVOIR RESORT

业主单位：所见西溪酒店管理有限公司
室内设计：卡纳设计
项目规模：40 000 m²
室内面积：40 000 m²
建成时间：2017年

Owner: Savoir Resort Management Co., Ltd.
Interior Design: CAC Design Group
Project Scale: 40,000 m²
Landscape Area: 40,000 m²
Date of Completion: 2017

Resort Hotel ·度假酒店·

项目回归空间本身，以中国传统文化"隐世哲学"为主轴，在中式的框架结构下革新，为所见西溪创造一种自由、开放，也更符合时代特性的新秩序。

Taking the traditional Chinese culture "Seclusive Philosophy" as the main axis, the project, innovated under the Chinese-style framework, returns to the space itself and creates a new order for Savoir Resort featured with the traits of freedom, openness and the times.

有朋青年共享社区（畅景公寓改造）
UPON COMMUNITY (CHANGJING APARTMENT RENOVATION DESIGN)

业主单位：天津滨海名苑投资有限公司
室内设计：筑土国际
项目规模：25 708 m²
室内面积：3 300 m²
建成时间：2017
设计周期：12个月

Owner: Tianjin Binhai Ming Yuan Investment Ltd.
Interior Design: Archiland
Project Scale: 25,708 m²
Interior Area: 3,300 m²
Date of Completion: 2017
Cycle of Design: 12 months

Renovation ·旧改·

设计师将提取的水细胞元素通过参数化处理形成泰森多边形的叶脉网络，对各区域进行有机整合。整个共享社区由不同形状的细胞模块组合而成，革新泳池空间面貌的同时增加各项功能模块。

The designers extract water cell elements from the original pool memory, and form Thiessen polygon through the parameterized processing, then combine the regional and functional organic unifies together. The entire shared community is made up by different shapes of cell module, which not only makes the pool space have a new look, also add the area of other functional modules.

郑州阳光城·檀悦售楼处
WAVING RIBBON

业主单位：阳光城集团
室内设计：KLID达观国际建筑设计事务所
项目规模：489 800 m²
室内面积：1 100 m²
建成时间：2017年
设计周期：3个月

Owner: Sunshine City Group Co., Ltd.
Interior Area: KLID (Kris Lin International Design)
Project Scale: 489,800 m²
Interior Area: 1,100 m²
Date of Completion: 2017
Cycle of Design: 3 months

Sales Center ·售楼中心·

设计师运用"舞动的丝带"进行空间界定，从地平面开始环绕包裹，构建出建筑内部的体量感与线条感，引导参观者游历内部空间。灰、白丝带既串联了一二楼所有空间，又对其做了区隔和界定。

The waving ribbon is used for the interior design as the boundary, which starts from the ground to wrap the building, creating a sense of volume and line, and guiding visitors to the interior space. The grey and white ribbon not only connects the 1st floor with the 2nd one, but also divides the space into different areas.

中鹰·黑森林 "多彩空间" 样板房
ZHONGYING BLACK FOREST 'CHROMATIC SPACES' SHOW APARTMENT

业主单位：上海中鹰置业有限公司
室内设计：IFG 伊波莱茨建筑设计
项目规模：250 m²
室内面积：250 m²
建成时间：2017年
设计周期：8个月

Owner: Eagle Real Estate Co., Ltd.
Interior Design: Ippolito Fleitz Group–Identity Architects
Project Scale: 250 m²
Interior Area: 250 m²
Date of Completion: 2017
Cycle of Design: 8 months

Model Room · 样板房 ·

"多彩空间"鲜明的色彩设计结合几何图形展现公寓活力，丰富的色彩拼接反映生活乐趣和放松的居住氛围。Loft风格的客厅和窗外美景相结合，使休息空间与大自然完美地融于一体，时刻激发住客的想象力。

"Chromatic Spaces" bright color design combined with geometric figures shows the vitality of the apartment, and the rich color splicing reflects the fun of life and the relaxing living atmosphere. The combination of the loft-like atmosphere and breath-taking views inspire the visitor's imagination with a well-balanced spirit of vibrant energy and calm naturalness.

广州金地壹阅府
GUANGZHOU GEMDALE THE ONE VILLA

业主单位：金地集团广州公司
室内设计：LSD Interior Design
建成时间：2018年
设计周期：4个月

Owner: Gemdale
Interior Design: LSD Interior Design
Date of Completion: 2018
Cycle of Design: 4 months

Model Room · 样板房 ·

项目对建筑内部空间进行颠覆性改动，创造更流畅的动线和更合理的空间资源分配，追求材料的真实、功能的释放和表达的克制，在这之上，再通过变化去创造一些日常中的小小趣味。

The project makes subversive changes to the internal space of the building, to create smoother generatrixes and more reasonable allocation of space resources. Besides, it pursues the genuineness of materials and the release of functions, and moderation of express on top of this, create some little surprises in everyday life through changes.

十字轴线之间
CHARMING EASTERN

业主单位：国美建设股份有限公司
室内设计：惹雅国际设计事务所
室内面积：380 m²
建成时间：2018年
设计周期：2.75个月

Owner: GOODMAN GROUP
Interior Design: THEE DESIGN
Interior Area: 380 m²
Date of Completion: 2018
Cycle of Design: 2.75 months

Private House · 私宅 ·

本项目打破传统住宅格局，以十字形轴线将走道一分为二，划分公私领域的隐性界线；以独特的切割方式铺陈整体空间，营造出极具艺术性的独特韵味。

Breaking the traditional residential structure, the aisle is divided into two by a cross-shaped axis, and the hid-den boundary between the public and private areas is divided. The project lays out the overall space with a unique cutting method to create a unique and artistic atmosphere.

Retail · 零售 ·

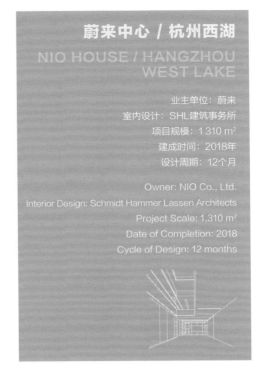

蔚来中心 / 杭州西湖
NIO HOUSE / HANGZHOU WEST LAKE

业主单位：蔚来
室内设计：SHL建筑事务所
项目规模：1 310 m²
建成时间：2018年
设计周期：12个月

Owner: NIO Co., Ltd.
Interior Design: Schmidt Hammer Lassen Architects
Project Scale: 1,310 m²
Date of Completion: 2018
Cycle of Design: 12 months

杭州西湖蔚来中心拥有俯瞰西湖的景致，是一处结合了展示空间和会员俱乐部的新型汽车展厅空间。蔚来品牌精髓与西湖文化融合在一起，让用户们可以多感官深入体验NIO蔚来向自然致敬的品牌哲学。

Located on the scenic West Lake waterfront in Hangzhou, China, the new flagship automotive gallery and clubhouse for car manufacturer, NIO blends the company brand with West Lake culture, giving consumers a multi-sensory understanding of NIO's environmentally-rooted brand philosophy.

伍兹贝格上海工作室
WOODS BAGOT SHANGHAI STUDIO

业主单位：伍兹贝格
室内设计：Woods Bagot 伍兹贝格
室内面积：1 300 m²
建成时间：2018年
设计周期：3个月

Owner: Woods Bagot
Interior Design: Woods Bagot
Landscape Area: 1,300 m²
Date of Completion: 2018
Cycle of Design: 3 months

Office Building · 办公项目 ·

项目整体设计融合生活气息与办公氛围，"中轴线"贯穿工作室，连接各个区域，吧台区可以灵活地调整为活动场地，灵活机动的空间布局在展示形态的同时更是从人性化的角度出发，促进团队间的互动与协作。

With the air of a contemporary hospitality space, Woods Bagot's Shanghai Studio is an embracing space for living and working. A central avenue that runs through the office to each area. The flexible bar area is a dynamic, flexible space that prioritizes the human experience, encouraging connections, fueling collaboration and ultimately expressing what Woods Bagot is truly about.

Private House · 私 宅 ·

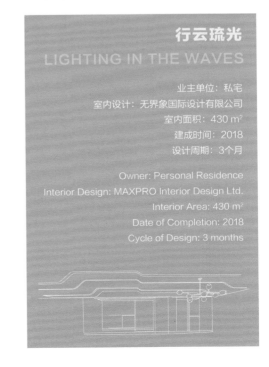

行云琉光
LIGHTING IN THE WAVES

业主单位：私宅
室内设计：无界象国际设计有限公司
室内面积：430 m²
建成时间：2018
设计周期：3个月

Owner: Personal Residence
Interior Design: MAXPRO Interior Design Ltd.
Interior Area: 430 m²
Date of Completion: 2018
Cycle of Design: 3 months

项目为纵深17米的高楼层住家，设计师利用自然景致创造"云中之家"，以居室意象整合三代同堂的空间需求。设计师拿捏精准比例，叠加空间感，让一室空间得以串联，如行云流水，自然流畅。

The project is a high-floor residence with a depth of 17 meters. The designer uses the natural scenery to create a "home in the cloud" and integrates the spatial needs of three generations with the image of the room. The precise proportion of the design is superimposed on the sense of space. Making a comfortable living space for the pleasure of the inhabitants is the most valuable undertaking as designers.

05 照 明 设 计 LIGHTING DESIGN

第四届金奖

4th GOLD AWARD

华侨城苏河湾上海宝格丽酒店
OCT SUHE CREEK BVLGARI HOTEL

苏悦广场
THE SUMMIT

第五届金奖

5th GOLD AWARD

长沙梅溪湖国际文化艺术中心
MEIXI LAKE INTERNATIONAL CULTURE ART CENTER

外滩金融中心
THE BUND FINANCE CENTER

郑州万科美景·世玠
VANKE MEIJING SHIJIE

第四届银奖

4th
SILVER
AWARD

乌镇阿丽拉酒店
WUZHEN ALILA HOTEL

第五届银奖

5th
SILVER
AWARD

华侨城苏河湾上海宝格丽酒店
OCT SUHE CREEK BVLGARI HOTEL

业主单位：华侨城（上海）置地有限公司
照明设计：Project Lighting Design Pte Ltd.
项目规模：234 091 m²
建成时间：2017年
设计周期：60个月

Owner: OCT Land (Shanghai) Investment Ltd.
Lighting Design: Project Lighting Design Pte Ltd.
Project Scale: 234,091 m²
Date of Completion: 2017
Cycle of Design: 60 months

第四届·金奖
4th GOLD AWARD

Hotel ·酒店·

华侨城苏河湾东区项目集商业、酒店、住宅、办公、历史建筑于一体，创造了苏州河沿岸优美、舒适、和谐的建筑形态以及酒店、办公区、商业区、住宅和谐共生的大环境。

The project of OCT Suhe Creek East sets commercial, hotel, residence office, historic buildings in one. It creates a beautiful, comfortable, harmonious architectural form and harmony environment with hotel, office, business area, commercial area and residential area along the Suzhou River.

第四届·金奖
4th GOLD AWARD

Commercial Project ·商办项目·

苏悦广场
THE SUMMIT

业主单位：铁狮门
照明设计：RDI瑞国际照明设计
项目规模：15 392 m²
建筑面积：149 735 m²
建成时间：2017年
设计周期：36个月

Owner: TISHMAN SPEYER
Lighting Design: RDESIGN INTERNATIONAL LIGHTING
Project Scale: 15,392 m²
Floor Area: 149,735 m²
Date of Completion: 2017
Cycle of Design: 36 months

作为一个商业综合体，灯光强调了一系列交错的体量，在视觉上将两个国际写字楼与高端住宅连接起来。设计灵感来源于苏州本土著名建筑形式的黑白灰色板，其被作为灯光的塑造语言。

As a commercial mix-used complex, the lighting emphasizes a series of interlocking volumes, and visually unifies the two parcels at night. The white, gray and black color palette inspired by the architectural vernacular famous in Suzhou, is also the language in the light.

长沙梅溪湖国际文化艺术中心
MEIXI LAKE INTERNATIONAL CULTURE ART CENTER

业主单位：长沙梅溪湖实业有限公司
照明设计：bpi
建筑面积：124 940 m²
建成时间：2018年

Owner: Changsha Meixihu Industry Co., Ltd.
Lighting Design: bpi
Floor Area: 124,940 m²
Date of Completion: 2018

第五届 金奖
5th GOLD AWARD

Culture Project ·文化项目·

设计师创造适合的灯光，捕捉流动的轨迹，将照明灯具与室内空间进行完美融合。扎哈的设计赋予了整个建筑空间永恒流动的生命形态。

The designers create suitable lighting, capture the flowing trajectory and integrate the lighting fixtures with the indoor space perfectly. Zaha's design gives the entire architectural space an eternal and flowing life form.

外滩金融中心
THE BUND FINANCE CENTER

业主单位：上海证大外滩国际金融服务中心置业有限公司
照明设计：bpi
项目规模：420 000 m²
建筑面积：190 000 m²
建成时间：2017年
设计时间：60个月

Owner: Shanghai Zendai Bund Intl
    Finance Center Real Estate
Lighting Design: bpi
Project Scale: 420,000 m²
Floor Area: 190,000 m²
Date of Completion: 2017
Cycle of Design: 60 months

第四届·银奖
4th SILVER AWARD

Mixed-use ·综合体·

照明设计以"世界，重回外滩"为理念，突出建筑与城市之间的开拓精神与内敛气质，用光来延续浦东浦西的传承、现代与未来的对话与过渡。定制化的光与灯具和建筑完美汇聚，成为永恒的经典。

Based on the concept of "The world returns to the Bund", the lighting design, highlighting the pioneering spirit and introverted temperament between the architecture and the city, uses lights to carry the extension between Pudong and Puxi, as well as the dialogue and transition between the present and the future. In addition, the customized light and lamps perfectly melt into the building, making it classical eternity.

第四届·银奖
4th SILVER AWARD

Demonstration Area ·示范区·

郑州万科美景·世玠
VANKE MEIJING SHIJIE

业主单位：郑州万科房地产有限公司
照明设计：中奥光科（北京）国际照明设计（IDDI）
项目规模：21 680 m²
建筑面积：15 000 m²
建成时间：2017年
设计周期：6个月

Owner: Vanke
Lighting Design: IDDI Design Group, inc
Project Scale: 21,680 m²
Floor Area: 15,000 m²
Date of Completion: 2017
Cycle of Design: 6 months

郑州纺织机床厂始于20世纪50年代，此次与城市一同再生长。在回顾景观界面后，灯光设计师决定延乘其古典美学，将光影定义为"光的咏叹调，影的叙事曲"，灯光设计的脉络也随之谱现。

Built in the 1950s, Zhengzhou Textile Machine Tool Plant is beating with the city. After reviewing the landscape interface, the designer decides to carry the classical aesthetics of the lighting design, defining light and shadow as "an aria of light, narrative of shadow", revealing the context of lighting as well.

乌镇阿丽拉酒店
WUZHEN ALILA HOTEL

业主单位：浙江雅达酒店投资管理有限公司
照明设计：bpi
项目规模：26 237 m²
建成时间：2018年
设计周期：22个月

Owner: Zhejiang Yada Hotel
    Investment Management Co., Ltd.
Lighting Design: bpi
Project Scale: 26,237 m²
Date of Completion: 2018
Cycle of Design: 22 months

第五届·银奖
5th SILVER AWARD

Resort Hotel ·度假酒店·

小桥流水与街巷的宁静赋予照明设计无限灵感，简约舒适的照明效果极配合当地建筑风格，将江南古民居的极简语言发挥到极致。宜人的照明尺度在保证防眩光控制的同时将能耗控制到最低。

The tranquil charm of small bridges and grunge streets gives Alila unlimited inspiration. Simple and comfortable lighting effect matches the minimalist architectural style. Because of the pleasant lighting scale, the garden lights are controlled under the line of sight and achieve anti-glare control. At the same time, the energy consumption of the whole project is controlled to the minimum.

评委名单
JURYS LIST

第一届地产设计大奖·中国 评委名单
1ST CREDAWARD JURYS LIST

主　席	邢同和	上海现代建筑设计集团 同济大学	资深总建筑师 教授、博士生导师
学术评委	李振宇	同济大学	建筑与城市规划学院院长
	孟建民	深圳市建筑设计研究总院有限公司 中国建筑学会 华南理工大学 东南大学	总建筑师 常务理事 硕士生导师 博士生导师
	庄惟敏	清华大学 清华大学建筑设计研究院有限公司	建筑学院院长、教授、博士生导师 院长、总建筑师
地产评委	陈建邦	瑞安房地产发展有限公司	规划发展及设计总监
	常　宇	祥源控股集团股份有限公司	副总裁
	陈栋梁	上海朗诗建筑科技有限公司	总经理
	范　炜	上海保利房地产开发有限公司	副总经理
	冯腾飞	万达集团	万达商业规划研究院副院长
	高　峰	仁恒置地集团	总建筑师兼总监
	郭咏海	亿达集团	总规划师
	何定康	新世界中国地产有限公司	标准化设计研发高级经理
	胡树志	招商局地产控股股份有限公司	副总建筑师
	贾朝晖	绿地集团	副总建筑师
	居培成	万科集团	上海区域总建筑师
	钱　毅	雨润控股集团	副总裁
	佘啸吟	碧桂园集团	定位策划部总经理
	寿　东	宝龙集团	研发定位中心总经理
	许海东	正大新生活	设计管理部助理副总裁
	阴　杰	SOHO中国	首席建筑师、高级副总裁
	于　鹏	世茂房地产控股有限公司	研发设计总监
	余章荣	联实集团	中国区发展及策划部总经理

	Name	Affiliation	Title
CHAIRMAN	Tonghe Xing	Shanghai Xian Dai Architectural Design (Group)Co.,Ltd Tongji University	Senior Chief Architect Professor & Doctoral Supervisor
ACADEMIC JURY	Zhenyu Li	Tongji University	Dean, College of Architecture and Urban Planning
	Jianmin Meng	SZAD Architectural Society of China South China University of Technology Southeast University	Chief Architect Execuive Director Graduate Supervisor Doctoral Supervisor
	Weimin Zhuang	Tsinghua University Architectural Design & Research Institute Tsinghua University Co.,Ltd.	Dean, School of Architecture, Professor, Doctoral Supervisor Dean & Chief Arhictect
DEVELOPER JURY	K.B. Albert Chan	Shui On Land	Director of Development Planning & Design
	Yu Chang	Sunriver Holding Group Co.,Ltd.	Vice President
	Dongliang Chen	Shanghai Landsea Architectural Technology Co.,Ltd.	General Manager
	Wei Fan	Poly Shanghai Real Estate Development Co.,Ltd.	Deputy General Manager
	Tengfei Feng	Wanda Group	Vice President of Wanda Commercial Planning Institute
	Feng Gao	Yanlord Land Group	Chief Architect & Director
	Yonghai Guo	Yida Group	Chief Planner
	Wallace Ho	New World China	Senior Manager of Standardized Design and Research
	Shuzhi Hu	China Merchants Property Development Co.,Ltd.	Deputy Chief Architect
	Zhaohui Jia	Greenland Group	Deputy Chief Architect
	Peicheng Ju	Vanke Group	Chief Architect of Shanghai Branch
	Yi Qian	YuRun Holdings Group Co.,Ltd.	Vice President
	Xiaoyin She	Country Garden Group	General Manager of Research Center
	Dong Shou	Powerlong Group	General Manager of Research Center
	Haidong Xu	Chia Tai Group	Assistant Vice President of Design Management Dept.
	Jerry Yin	SOHO China	Chief Architect, Senior Vice President
	Steven Yu	Shimao Property Holdings Ltd.	Director of Research & Design Dept.
	Howard Zhangrong Yu	Lend Lease Group	Head of Development & Planning

第二届地产设计大奖·中国 评委名单
2ND CREDAWARD JURYS LIST

主　席	邢同和	上海现代建筑设计集团 同济大学	资深总建筑师 教授、博士生导师
学术评委	孟建民	深圳市建筑设计研究总院 中国工程院 中国建筑学会 华南理工大学 东南大学	总建筑师 院士 常务理事 硕士生导师 博士生导师
	王建国	东南大学 中国工程院 中国城市规划学会 中国建筑学会 住房和城乡建设部	教授 院士 常务理事兼城市设计学术委员会副主任 常务理事 城乡规划专家委员会委员
地产评委	陈建邦	瑞安房地产发展有限公司	规划发展及设计总监
	常　宇	祥源控股集团股份有限公司	副总裁
	陈栋梁	上海朗诗建筑科技有限公司	总经理
	范　炜	上海保利房地产开发有限公司	副总经理
	冯腾飞	万达集团	万达商业规划研究院副院长
	高　峰	仁恒置地集团	总建筑师兼总监
	郭咏海	新加坡星桥腾飞集团	地产开发部助理副总裁
	何定康	新世界中国地产有限公司	标准化设计研发高级经理
	胡树志	招商蛇口	副总建筑师
	贾朝晖	绿地集团	副总建筑师
	居培成	万科集团	上海区域总建筑师
	钱　毅	雨润控股集团	副总裁
	佘啸吟	碧桂园集团	定位策划部总经理
	寿　东	宝龙集团	研发定位中心总经理
	许海东	正大新生活	设计管理部助理副总裁
	杨　凡	华侨城（上海）置地有限公司	副总经理
	阴　杰	SOHO中国	首席建筑师、高级副总裁
	于　鹏	世茂房地产控股有限公司	研发设计总监
	余章荣	联实集团	中国区发展及策划部总经理
	朱晓涓	上海证大房地产	副总裁

CHAIRMAN	Tonghe Xing	Shanghai Xian Dai Architectural Design (Group)Co.,Ltd Tongji University	Senior Chief Architect Professor & Doctoral Supervisor
ACADEMIC JURY	Jianmin Meng	SZAD Chinese Academy of Engineering Architectural Society of China South China University of Technology Southeast University	Chief Architect Academician Execuive Director Graduate Supervisor Doctoral Supervisor
	Jianguo Wang	Southeast University Chinese Academy of Engineering Urban Planning Society of China Architectural Society of China Ministry of Housing and Urban-Rural Development	Professor Academician Execuive Director & Deputy Director, Academic Committee of Urban Design Execuive Director Member, Expert Committee of Urban and Rural Planning
DEVELOPER JURY	K.B. Albert Chan	Shui On Land	Director of Development Planning & Design
	Yu Chang	Sunriver Holding Group Co.,Ltd.	Vice President
	Dongliang Chen	Shanghai Landleaf Architectural Technology Co.,Ltd.	General Manager
	Wei Fan	Poly Shanghai Real Estate Development Co.,Ltd.	Deputy General Manager
	Tengfei Feng	Wanda Group	Vice President of Wanda Commercial Planning Institute
	Feng Gao	Yanlord Land Group	Chief Architect & Director
	Yonghai Guo	Ascendas-Singbridge	Assistant Vice President of REDS
	Wallace Ho	New World China	Deputy General Manager of Standardized Design and Research
	Shuzhi Hu	China Merchants Shekou Holdings	Deputy Chief Architect
	Zhaohui Jia	Greenland Group	Deputy Chief Architect
	Peicheng Ju	Vanke Group	Chief Architect of Shanghai Branch
	Yi Qian	YuRun Holdings Group Co.,Ltd.	Vice President
	Xiaoyin She	Country Garden Group	General Manager of Research Center
	Dong Shou	Powerlong Group	General Manager of Research Center
	Haidong Xu	Chia Tai Group	Assistant Vice President of Design Management Dept.
	Fan Yang	OCT Land (shanghai) Investment Ltd.	Deputy General Manager
	Jerry Yin	SOHO China	Chief Architect, Senior Vice President
	Steven Yu	Shimao Property Holdings Ltd.	Director of Research & Design Dept.
	Howard Zhangrong Yu	Lend Lease Group	Head of Development & Planning
	Xiaojuan Zhu	Shanghai Zendai Property Co.,Ltd.	Vice President

第三届地产设计大奖·中国 评委名单
3RD CREDAWARD JURYS LIST

主 席	邢同和	上海现代建筑设计集团 同济大学	资深总建筑师 教授、博士生导师
学术评委	孟建民	深圳市建筑设计研究总院 中国工程院 中国建筑学会 华南理工大学 东南大学	总建筑师 院士 常务理事 硕士生导师 博士生导师
	王建国	东南大学 中国工程院 中国城市规划学会 中国建筑学会 住房和城乡建设部	教授 院士 常务理事兼城市设计学术委员会副主任 常务理事 城乡规划专家委员会委员
	王 浩	南京林业大学 中国林业教育学会 中国风景园林学会 住房和城乡建设部 国家湿地科学技术	校长、博士生导师 副会长 常务理事 风景园林专家委员会委员 专家委员会委员
地产评委	陈建邦	瑞安房地产发展有限公司	规划发展及设计总监
	常 宇	祥源控股集团股份有限公司	副总裁
	陈栋梁	上海朗绿科技股份有限公司	总裁,合伙人
	陈海亮	万达集团	万达商业规划研究院总建筑师
	范逸汀	旭辉控股(集团)有限公司	副总裁
	冯腾飞	中弘控股股份有限公司	旅游规划研究院院长
	高 峰	仁恒置地集团	总建筑师兼总监
	郭咏海	新加坡星桥腾飞集团	地产开发部助理副总裁
	胡树志	招商蛇口	副总建筑师
	胡 浩	龙湖集团	总建筑师、研发部总经理
	黄宇奘	碧桂园集团	副总裁、总设计师
	贾朝晖	绿地集团	副总建筑师
	居培成	万科集团	上海区域总建筑师
	钱 毅	雨润控股集团	党委书记、副总裁
	佘啸吟	蓝光集团	广佛公司总经理
	寿 东	鲁能集团	杭州区域设计总监
	杨 凡	华侨城(上海)置地有限公司	副总经理
	阴 杰	SOHO中国	首席建筑师、高级副总裁
	于 鹏	融信中国	设计管理中心总经理
	张兆强	阳光城集团	产品研发中心总经理
	朱晓涓	上海证大房地产	副总裁

CHAIRMAN	Tonghe Xing	Shanghai Xian Dai Architectural Design (Group)Co.,Ltd Tongji University	Senior Chief Architect Professor & Doctoral Supervisor
ACADEMIC JURY	Jianmin Meng	SZAD Chinese Academy of Engineering Architectural Society of China South China University of Technology Southeast University	Chief Architect Academician Execuive Director Graduate Supervisor Doctoral Supervisor
	Jianguo Wang	Southeast University Chinese Academy of Engineering Urban Planning Society of China Architectural Society of China Ministry of Housing and Urban-Rural Development	Professor Academician Execuive Director & Deputy Director, Academic Committee of Urban Design Execuive Director Member, Expert Committee of Urban and Rural Planning
	Hao Wang	Nanjing Forestry University China Education Association of Forestry (CEAF) Chinese Society of Landscape Architecture (CHSLA) Housing & Urban-Rural Development National Wetland Science & Technology	Principal, Doctoral Supervisor Vice President Execuive Director Expert Committee Member of Landscape Architecture Expert Committee Member
DEVELOPER JURY	K.B. Albert Chan	Shui On Land	Director of Development Planning & Design
	Yu Chang	Sunriver Holding Group Co.,Ltd.	Vice President
	Dongliang Chen	Shanghai Landleaf Architectural Technology Co.,Ltd.	President ,Partner
	William Hailiang Chen	Wanda Group	Chief Architect of Wanda Commercial Planning Institute
	Yiting Fan	CIFI Holdings (Group) Co.,Ltd.	Vice President
	Tengfei Feng	Zhonghong Holding Co.,Ltd.	Director of Tourism Plannig & Research Institute
	Feng Gao	Yanlord Land Group	Chief Architect & Director
	Yonghai Guo	Ascendas-Singbridge	Assistant Vice President of REDS
	Shuzhi Hu	China Merchants Shekou Holdings	Deputy Chief Architect
	Hao Hu	Longfor Group	Chief Architect,General Manager of Research & Development Dept.
	Yuzang Huang	Country Garden Group	Vice President ,Chief Designer
	Zhaohui Jia	Greenland Group	Deputy Chief Architect
	Peicheng Ju	Vanke Group	Chief Architect of Shanghai Branch
	Yi Qian	YuRun Holdings Group Co.,Ltd.	Vice President
	Xiaoyin She	BRC Group	General Manager of Research Center
	Dong Shou	Luneng Group	Design Director of Hangzhou Branch
	Fan Yang	OCT Land (shanghai) Investment Ltd.	Deputy General Manager
	Jerry Yin	SOHO China	Chief Architect,Senior Vice President
	Steven Yu	Ronshine China	General Manager of Design Administration Center
	Zhaoqiang Zhang	Shanghai Zendai Property Co.,Ltd.	Vice President
	Xiaojuan Zhu	Yango Group	General Manager of Product R&D Center

第四届地产设计大奖·中国 评委名单
4TH CREDAWARD JURYS LIST

主席	邢同和	上海现代建筑设计集团 同济大学	资深总建筑师 教授、博士生导师
学术评委	孟建民	深圳市建筑设计研究总院 中国工程院 中国建筑学会 华南理工大学 东南大学	总建筑师 院士 常务理事 硕士生导师 博士生导师
	王建国	东南大学 中国工程院 中国城市规划学会 中国建筑学会 住房和城乡建设部	教授 院士 常务理事兼城市设计学术委员会副主任 常务理事 城乡规划专家委员会委员
	王浩	南京林业大学 中国林业教育学会 中国风景园林学会 住房和城乡建设部 国家湿地科学技术	校长、博士生导师 副会长 常务理事 风景园林专家委员会委员 专家委员会委员
地产评委	陈建邦	瑞安房地产发展有限公司	规划发展及设计总监
	常宇	祥源控股集团股份有限公司	副总裁
	陈栋梁	上海朗绿科技股份有限公司	总裁、合伙人
	范炜	陕西保利房地产开发有限公司	总经理
	范逸汀	旭辉控股（集团）有限公司	副总裁
	方芳	万达集团	万达商业规划研究院副总规划师
	冯腾飞	山水文园集团	策研设计总院联席院长
	高峰	融创中国	北京区域集团研发中心总经理
	郭咏海	协信控股（集团）有限公司	小镇事业部助理总裁
	胡树志	招商局地产控股股份有限公司	副总建筑师
	胡浩	龙湖集团	副总裁、研发部总经理
	黄宇奘	碧桂园集团	副总裁、总设计师
	贾朝晖	绿地集团	江苏房地产事业部副总经理
	居培成	万科集团	上海区域总建筑师
	钱毅	雨润控股集团	党委书记、副总裁
	佘啸吟	方直集团	副总裁
	寿东	鲁能集团	杭州区域设计总监
	杨凡	华侨城（上海）置地有限公司	副总经理
	阴杰	铁狮门	董事总经理，设计与工程
	于鹏	中国金茂	产品&工程管理中心总经理
	张春玲	复星地产	副总裁、合伙人
	张胜利	金辉集团	副总裁
	张兆强	阳光城集团	副总裁
	朱晓涓	上海证大房地产	副总裁

CHAIRMAN	Tonghe Xing	Shanghai Xian Dai Architectural Design (Group)Co.,Ltd Tongji University	Senior Chief Architect Professor & Doctoral Supervisor
ACADEMIC JURY	Jianmin Meng	SZAD Chinese Academy of Engineering Architectural Society of China South China University of Technology Southeast University	Chief Architect Academician Execuive Director Graduate Supervisor Doctoral Supervisor
	Jianguo Wang	Southeast University Chinese Academy of Engineering Urban Planning Society of China Architectural Society of China Ministry of Housing and Urban-Rural Development	Professor Academician Execuive Director & Deputy Director, Academic Committee of Urban Design Execuive Director Member, Expert Committee of Urban and Rural Planning
	Hao Wang	Nanjing Forestry University China Education Association of Forestry (CEAF) Chinese Society of Landscape Architecture (CHSLA) Housing & Urban-Rural Development National Wetland Science & Technology	Principal, Doctoral Supervisor Vice President Execuive Director Expert Committee Member of Landscape Architecture Expert Committee Member
DEVELOPER JURY	K.B. Albert Chan	Shui On Land	Director of Development Planning & Design
	Yu Chang	Sunriver Holding Group Co.,Ltd.	Vice President
	Dongliang Chen	Shanghai Landleaf Architectural Technology Co.,Ltd.	President, Partner
	Wei Fan	Poly Shaanxi Real Estate Development Co.,Ltd.	General Manager
	Yiting Fan	CIFI Holdings (Group) Co.,Ltd.	Vice President
	Fang Fang	Wanda Group	Deputy Chief Planner of Wanda Commercial Planning & Research Institute Co., Ltd.
	Tengfei Feng	Riverside Group	Joint President of Strategic Planning & Research Institute
	Feng Gao	Sunac Group	General Manager of R&D Cente,Beijing Branch
	Yonghai Guo	Sincere Group	Assistant President of Small Town Division
	Shuzhi Hu	China Merchants Property Development Co.,Ltd.	Deputy Chief Architect
	Hao Hu	Longfor Group	Vice President,General Manager of Research & Development Dept.
	Yuzang Huang	Country Garden Group	Vice President,Chief Designer
	Zhaohui Jia	Greenland Group	Deputy General Manger of Jiangsu Real Estate Business Unit
	Peicheng Ju	Vanke Group	Chief Architect of Shanghai Branch
	Yi Qian	YuRun Holdings Group Co.,Ltd.	Vice President
	Xiaoyin She	Faithland Group	Vice President
	Dong Shou	Luneng Group	Design Director of Hangzhou Branch
	Fan Yang	OCT Land (shanghai) Investment Ltd.	Deputy General Manager
	Jerry Yin	Tishman Speyer	Managing Director, Design & Construction
	Steven Yu	China Jinmao	General Manager of Product Administration Center
	Chunling Zhang	FOSUN Property	Vice President, Partner
	Shengli Zhang	Radiance Group	Vice President
	Zhaoqiang Zhang	Yango Group	Vice President
	Xiaojuan Zhu	Shanghai Zendai Property Co.,Ltd.	Vice President

第五届地产设计大奖·中国 评委名单
5TH CREDAWARD JURYS LIST

主 席	邢同和	上海现代建筑设计集团 同济大学	资深总建筑师 教授、博士生导师
学术评委	孟建民	深圳市建筑设计研究总院 中国工程院 中国建筑学会 华南理工大学 东南大学	总建筑师 院士 常务理事 硕士生导师 博士生导师
	王建国	东南大学 中国工程院 中国城市规划学会 中国建筑学会 住房和城乡建设部	教授 院士 常务理事兼城市设计学术委员会副主任 常务理事 城乡规划专家委员会委员
	Martha Thorne	西班牙 IE 建筑学院	院长
地产评委	陈建邦	瑞安房地产发展有限公司	规划发展及设计总监
	常 宇	祥源控股集团股份有限公司	副总裁
	陈栋梁	上海朗绿科技股份有限公司	总裁，董事，创始合伙人
	范 炜	陕西保利房地产开发有限公司	总经理
	范逸汀	旭辉控股（集团）有限公司	副总裁
	方 芳	万达集团	万达商业规划研究院副总规划师
	冯腾飞	富康实业集团	副总裁
	高 峰	融创中国	北京区域集团烟威区域副总经理
	胡树志	招商局地产控股股份有限公司	专家组副总建筑师
	胡 浩	中国奥园集团	高级副总裁
	黄宇奘	碧桂园集团	副总裁
	贾朝晖	绿地集团	江苏房地产事业部副总经理
	茅 勤	中南置地	研发设计中心总经理
	钱 毅	雨润控股集团	党委书记兼副总裁
	戎武杰	绿地集团	总建筑师，技发部总经理
	佘啸吟	蓝光集团	深圳区域董事长
	寿 东	鲁能集团	杭州区域设计总监
	黄天龙	香港置地集团	中国区物业发展及投资副董事兼设计主管
	许 洁	建发房产	总建筑师
	杨 凡	华侨城华东集团	总建筑师
	阴 杰	铁狮门	董事总经理，设计与工程
	张春玲	豫园股份 复地企业发展有限公司	合伙人、首席设计师 合伙人、高级副总裁
	张兆强	阳光城集团	副总裁、高级合伙人
	朱晓涓	上海证大房地产	副总裁
	祝 峥	金地集团	设计总监

CHAIRMAN	Tonghe Xing	Shanghai Xian Dai Architectural Design (Group)Co.,Ltd Tongji University	Senior Chief Architect Professor & Doctoral Supervisor
ACADEMIC JURY	Jianmin Meng	SZAD Chinese Academy of Engineering Architectural Society of China South China University of Technology Southeast University	Chief Architect Academician Execuive Director Graduate Supervisor Doctoral Supervisor
	Jianguo Wang	Southeast University Chinese Academy of Engineering Urban Planning Society of China Architectural Society of China Ministry of Housing and Urban-Rural Development	Professor Academician Execuive Director & Deputy Director, Academic Committee of Urban Design Execuive Director Member, Expert Committee of Urban and Rural Planning
	Martha Thorne	IE School of Architecture and Design	Dean
DEVELOPER JURY	K.B. Albert Chan	Shui On Land	Director of Development Planning & Design
	Yu Chang	Sunriver Holding Group Co.,Ltd.	Vice President
	Dongliang Chen	Shanghai Landleaf Architectural Technology Co.,Ltd.	President, Director, Founding Partner
	Wei Fan	Poly Shaanxi Real Estate Development Co.,Ltd.	General Manager
	Yiting Fan	CIFI Holdings (Group) Co.,Ltd.	Vice President
	Fang Fang	Wanda Group	Deputy Chief Panner of Wanda Commercial Planning Institute
	Tengfei Feng	Falcon Industrial Group	Vice President
	Feng Gao	Sunac Group	Vice General Manager of Beijing Regional Group Yanwei Company
	Shuzhi Hu	China Merchants Property Development Co.,Ltd.	Deputy Chief Architect of Expert Team
	Hao Hu	China Aoyuan Group	Senior Vice President
	Yuzang Huang	Country Garden Group	Vice President
	Zhaohui Jia	Greenland Group	Deputy General Manger of Jiangsu Real Estate Business Unit
	Jose Mao	Zoina Land	General Manager of R&D Center
	Yi Qian	YuRun Holdings Group Co.,Ltd.	Vice President
	Wujie Rong	Greenland Group	Chief Architect
	Xiaoyin She	BRC Group	Chairman of Shenzhen Branch
	Dong Shou	Luneng Group	Design Director of Hangzhou Branch
	Terence Wong	Hongkong Land	Associate Director & Head of Design, China Property
	Jie Xu	C&D Real Estate	Chief Architect
	Fan Yang	OCT Eastern China Investment Co.,Ltd.	Chief Architect
	Jerry Yin	Tishman Speyer	Managing Director, Design & Construction
	Chunling Zhang	YUYUAN INC. Shanghai Forte Enterprise Development Co.,Ltd.	Partner & Chief Design Officer Partner & Senior Vice President
	Zhaoqiang Zhang	Yango Group	Vice President, Senior Partner
	Xiaojuan Zhu	Shanghai Zendai Property Co.,Ltd.	Vice President
	Zheng Zhu	Gemdale Corporation	Design Director

评委寄语
MESSAGE FROM JURYS

· 标注：按姓名首字母排序
Sort by First Letter of Last Name

陈栋梁 Dongliang Chen

在解决居住的有无问题之后，我们对房子有了更多的期望。好的设计不仅是真、善、美的体现，更是地产成功经营的基石，还是我们对美好生活的规划！期望「CREDAWARD 地产设计大奖·中国」能引领地产行业的设计理念，让设计创造更多价值！

After solving the problem of having a shelter for us to live, we have more expectations for the houses. It's clear that a good design is not only a reflection of truth, kindness and beauty, but also a cornerstone of the successful operation of real estate, as well as, our plan for a better life! Thus, I hope the CREDAWARD will lead the design concept of the real estate industry and make design create more value!

陈建邦 K.B. Albert Chan

再次祝贺获奖者们！这本书集中呈现了中国地产设计领域最优秀的一批作品。它传达出一个重要的事实：好的设计能促进商业的成功，而好的创意更是能让两者相得益彰。环顾我们所生活的城市，大部分的设计都来自于地产开发项目。因此，要打造一个美好的、可持续发展的环境，关键就在于要做出好的地产设计。所以，恭喜本书的出版！这将为我们带来宝贵的经验总结。

I would like to express my congratulations to the winners again! This book intensively displays the most outstanding works in the field of real estate design in China, which conveys an important fact: A good design promotes a successful business, while a good idea makes both of them complement each other. Looking around the city where we live, we will find that the majority of the designs originate from real estate development projects. Therefore, it is key to create a good real estate design to make the environment friendly and sustainable. That's why I take the chance to congratulate the publication of this book which will bring us a valuable experience!

丁芳 Fang Ding

「CREDAWARD 地产设计大奖·中国」秉承共识的价值原则，公平公正地评价、褒奖并宣传那些顺应自热和人类本性，与其自身的即时环境相协调的可持续的"好设计"地产产品及创造它们的"产品人"。

The CREDAWARD adheres to the consensual principle of value as well as the fair and impartial evaluation. The CREDAWARD praises and popularizes the sustainable projects with good designs, which accord with their current environments and conform to nature and human. The CREDAWARD also honors those who create the projects.

常宇 Yu Chang

「CREDAWARD 地产设计大奖·中国」推动设计创新，引领设计潮流，树立设计典范！

The CREDAWARD promotes innovation, leads trends, and sets a paradigm in design！

范炜 Wei Fan

希望我们共同努力，多做点"建筑"，少盖点"房子"，一起促进地产行业健康发展！

I hope we will make concerted efforts to do more "architecture" and less "buildings" to jointly facilitate the healthy development of the real estate industry!

范逸汀 / Yiting Fan

近年来，中国房地产的规模不断扩张，2019年国内商品房市场规模达到了16万亿，这样的行业大背景催生出「CREDAWARD地产设计大奖·中国」的不断进化，并见证了这样一个产品设计领域不断突破的成长过程。随着行业更加成熟，更加回归产品价值，我相信会有更好的作品涌现出来，「CREDAWARD地产设计大奖·中国」的水准也会越来越高，希望我们一同书写新的历史。

In recent years, with the scale of the real estate continually expanding in China, the domestic commercial housing market reached 16 trillion RMB in 2019, which contributed to the improvement of CREDAWARD. It can be said that the CREDAWARD has witnessed the growing process of the ongoing breakthrough in the field of product design. As the real estate industry gets more mature and values its products, I believe the future works will be better, and the level of CREDAWARD will reach record high. I hope we will write a new history together.

高峰 / Feng Gao

这些年间，「CCREDAWARD地产设计大奖·中国」一直在"变"。每一年，作品的数量和质量、赛事的组织能力、评委构成都在不断提升。希望通过我们共同的努力，激励行业的不断创新，鼓励设计师不断推陈出新。不管行业如何起伏，我们希望我们所评审和倡导的设计带来的产品力永远是行业的催化剂。

The CREDAWARD has been improving through these years. Every year, the number of works is growing, the quality of works is being improved, the organizational ability is getting efficient, and more influential jurys are being involved in. We hope that through our joint efforts, we will inspire continuous innovation in the industry and encourage designers to constantly innovate. No matter how the industry rises and falls, we hope that the product power brought by the design we have reviewed and advocated will always be the catalyst of the industry.

方芳 / Fang Fang

我们希望与地产、设计行业里的各方力量一起，将当下有些单调的城市重新塑造成更具个性和美学吸引力的生活空间。也许我们做的「CREDAWARD地产设计大奖·中国」就像"一石投湖"，但我相信越来越多的人会看到涟漪中的美丽。

We hope to work with all parties in the real estate and design industries to reshape the present monotonous citiy into a more personal and aesthetically attractive living space. Perhaps the CREDAWARD is like "throwing one stone into a lake", which symbolizes it will have a ripple effect on the real estate and design industry. But I believe more and more people will see the beauty in the ripples.

郭咏海 / Yonghai Guo

「CREDAWARD地产设计大奖·中国」在充分对接市场与客户需求的前提下，利用互联网的手段让专业评委与公众一起参与到评选的过程中来，是一次极其有益的创新。它不仅集中展现了优秀的作品，推广了品牌，获奖作品还将建立一种价值观和秩序，引导建筑设计与地产开发迈入更加科学、更人性化及更可持续的新阶段。

Under the premise of fully matching the market and customer needs, the CREDAWARD uses the Internet to let the professional jurys and the public jointly participate in the selection process, which is a positive and innovative. Not only does the CREDAWARD focus on showing the outstanding works and promoting the brands, but also the winning works will establish the value and order, and lead the architectural design and real estate development into a more scientific, more humane and more sustainable new stage.

冯腾飞 / Tengfei Feng

一路走来，有说不完的钩沉往事，有看得见的累累硕果，有挥不去的情愫缱绻。不管世事如何变幻，只要行业还在，地产设计就在，「CREDAWARD地产设计大奖·中国」就要心怀执着，追求不舍，砥砺奋进，共赢未来。

Looking back, there are not only countless lessons and experience but also many visible achievements, which are lingering on my mind. No matter how things change in the world, as long as the industry is still there, the real estate design is there. The CREDAWARD will hold the persistent spirit, forge ahead and win a win-win future.

何定康 / Wallance Ho

「CREDAWARD地产设计大奖·中国」在短短的6年间，凭着专业、认真、公平及公正的评审理念，已经成为房地产开发及地产设计行业热切希望获得的殊荣之一。祝贺本书上各个得奖项目及团队！

After the brief 6 years, the CREDAWARD has secured a reputation of being conscientious, professional and fair-minded. As a result, the CREDAWARD has become one of the most coveted awards for real estate's developers and the design services industry. Congratulations to all the projects and teams on this catalog !

胡 浩 Hao Hu

感谢全体行业同人对本书出版给予的多方支持！多年来，「CREDAWARD 地产设计大奖·中国」以其公正专业的奖项评审，推动整个行业的产品设计朝着更加卓越的方向不断前行。时至今日，产品力早已成为地产企业毋庸置疑的核心竞争力。尤其是在后疫情时代，整个行业都对产品进行了反思与升级，更加关注居住环境的人性化、回归健康的生活方式。期待我们在CREDAWARD 地产设计大奖的舞台上，不断涌现优秀的产品，我们共同期待未来的城市生活一定会更加健康、更加美好！

I would like to convey my sincere thanks to all my colleagues for your support in publishing this book! For many years, The CREDAWARD, based on its impartial and professional judging awards, has driven the product design in the entire industry towards the upper level. Up to now, the product power has become the undoubted core competitiveness of the real estate companies. Especially in the post-epidemic era, the entire industry has reflected and upgraded its products, paying more attention to the humanization of the living environments and returning to a healthy lifestyle. We expect the excellent products will continuously emerge from the stage of CREDAWARD, and we also look forward to enjoying the better and healthier cities in the future.

黄天龙 Terence Wong

「CREDAWARD 地产设计大奖·中国」是非常有意义的一个奖项活动，我们关注市场的变化，对应市场的需求，每年度组织评审筛选行业内的优秀作品并颁发奖项。我们致力于推动中国地产业更卓越健康、更高水平的发展方向，同时肩负宣传中国地产界和设计界的责任，组织对外国际交流和对内跨界跨专业交流分享，促进行业内的互动交流和技术创新分享，共同使中国建筑更国际化，成为更能引领时尚、文化体验的艺术载体。

The CREDAWARD is meaningful. We focus on market changes and respond to market needs. We annually organize to review and screen the outstanding works in real estate industry and award them. We are committed to promoting the better, higher and healthier development of real estate industry in China. At the same time, we shoulder the responsibility of publicizing the Chinese real estate industry and the Chinese design community, organize international communication and cross-professional ideas exchange at homeland, and promote interaction, communication and sharing technological innovation within the industry, jointly facilitate Chinese architecture to go to the world, to lead the fashion and to let the world taste Chinese culture.

黄宇奘 Yuzang Huang

「CREDAWARD 地产设计大奖·中国」始终秉持客观公正专业的评审态度和行业公信力，以积极开放的心态吸引海内外建筑师/设计师在大奖的平台上展示卓越的设计创新力和对未来趋势的探索！期待「CREDAWARD 地产设计大奖·中国」，在全体地产界评委及建筑师/设计师的共同努力下，真正成为中国地区最具号召力和影响力的大奖，成为未来的普利兹克奖！

With an active and open mind, the CREDAWARD always upholds an objective and impartial professional judging attitude and industry credibility to attract the architects/designers at home and abroad to demonstrate the excellent design innovation and exploration of future trends on this platform! I expect the CREDAWARD, together with the joint efforts of all the jurys and architects/designers from the real estate industry, will truly become the most appealing and influential award in China and have the same influence as The Pritzker Architecture Prize does in the future!

贾朝晖 Zhaohui Jia

我们邀请全球范围内优秀的项目在中国展现，世界设计精英汇聚「CREDAWARD 地产设计大奖·中国」。希望我们共同助推中国设计强国梦，让中国建筑与世界交流，让中国文化立足世界舞台！

We invite the world-wide outstanding projects to be demonstrated in China, letting the design elites from the world gather at the CREDAWARD, I hope we will jointly make Chinese design excel in the world, letting Chinese architecture communicate with the world and Chinese culture have a seat in the world stage.

居培成 Peicheng Ju

「CREDAWARD 地产设计大奖·中国」以地产行业的综合视角来看待当前中国的设计在城市、环境、社区营造、景观、生态、建筑、室内、专项等方向上，在城市、环境和客户综合维度的项目上的价值创造，也希望借此评奖来建立关于地产行业好设计的标准。

The CREDAWARD takes a comprehensive perspective from the real estate industry to view the creation values of present Chinese design covering cities, environments, communities, landscapes, ecologies, architectures, indoors, special projects, etc, which are embodied in the projects with integrated dimensions of cities, environments and customers. I hope to use this award to establish standards for good design in the real estate industry.

茅 勤 / Jose Mao

地产设计这些年，平庸的多，创见的少。换个角度思考，在中国这种急匆匆的市场氛围下，设计创造价值的想象空间还很大。「CREDAWARD 地产设计大奖·中国」正是提供了一个非常优质的平台，祝越办越好！

Through those years, there have been many ordinary works but a few of creative ones in the real estate design. In other words, there is still a lot of room for us to create values in the design industry in China where the businesses are growing rapidly. It is the CREDAWARD that provides a very high-quality platform for us. I wish the CREDAWARD better and better!

戎武杰 / Wujie Rong

作为建筑师身份时，我享受设计给人带来美好改变的价值，当然有时只是梦想；作为地产建筑师时，给满怀匠心的建筑师、设计师带来实现梦想的信心，仍然让我享受；「CREDAWARD 地产设计大奖·中国」就是这样的璀璨舞台！

As an architect, I enjoy the value that design brings beautiful changes to people, although it is true that sometimes a design is just a dream. As a real estate architect, I still enjoy the moment when I bring the confidence into the architects and designers who are full of ingenuity to realize their dreams. To my knowledge, I think the CREDAWARD is a shining stage!

彭 冲 / Chong Peng

甄选优秀地产项目，推动住宅体系创新，引领未来人居方式！

Select outstanding real estate projects, promote the innovation of the housing system, and lead the way of future living!

佘啸吟 / Xiaoying She

愿「CREDAWARD 地产设计大奖·中国」成为中国地产奖项中的"航母"，中国设计奖项中的"普利兹克"！

May the CREDAWARD become the "aircraft carrier" in the Chinese real estate awards and "The Pritzker Architecture Prize" in the Chinese design awards!

钱 毅 / Yi Qian

希望通过「CREDAWARD 地产设计大奖·中国」进一步融汇中西文化、思想，弘扬传统文化，促进中国房地产产业的健康发展，扩大业界的交流，积极引导房地产产品的主流发展方向，为中国百姓能住上环保、高品质的住宅和享受美好生活而贡献一份力量。

I hope that through the CREDAWARD, we will further integrate Chinese cultures and ideas into Western ones, carry forward the traditional Chinese culture, promote the healthy development of real estate industry in China, expand communicating scope in the professional fields, and actively guide the direction of mainstream development of real estate products, contribute a positive effort for Chinese people to live in eco-friendly and high-quality houses and to enjoy a better life.

寿 东 / Dong Shou

「CREDAWARD 地产设计大奖·中国」将是国内外瞩目的一流奖项，希望大奖能够越走越高，涵盖更多的国内外优秀作品，通过设计连接中国与世界！

The CREDAWARD will be a first-class award that attracts attention domestically and globally in the future. I hope that the CREDAWARD will continuously reach a higher level, cover more outstanding works from home and abroad, and link China and the world with design!

田九坡 Jone Tian

祝「CREDAWARD 地产设计大奖·中国」越办越好，成为地产设计的风向标。

May the CREDAWARD get better and become the benchmark of real estate design.

杨 凡 Fan Yang

「CREDAWARD 地产设计大奖·中国」是联通地产和设计行业的桥梁，更是两个行业融合和交流的重要平台。愿大奖有着更广泛的影响力，更专业的判断力，更长久的生命力。

The CREDAWARD is not only a bridge to connect the real estate industry with the design industry but also an important platform for them to integrate and communicate. May the CREDAWARD have wider influence, more professional judgment, and longer vitality.

冼耀强 Yaoqiang Xian

非常荣幸能受邀参与到「CREDAWARD 地产设计大奖·中国」评委会的工作，这是一个让地产人备感兴奋的奖项，一个能让好项目好设计发光发热的平台，希望我们能在这次盛会上共同发现行业最闪亮的项目、最具引领性的设计。

It is a great honor to be invited to be the jury of the CREDAWARD, which is an exciting award for everyone in real estate field and is a platform where excellent projects and good designs can glow. I hope we will find the most outstanding projects and the leading designs in the CREDAWARD.

阴 杰 Jerry Yin

非常荣幸参与到「CREDAWARD 地产设计大奖·中国」的评委会的工作中来，让我们一起用心发现属于中国的好建筑，代表世界水准的好设计！希望更多优秀作品、杰出的设计公司参与到奖项中来！

It is my honor to be a part of the CREDAWARD Jury! Let us devote to discovering a good building that belongs to China and represents a world-class design! I also hope more excelled works and outstanding design companies will participate in the CREDAWARD!

许 洁 Jie Xu

一个百花齐放的时代，需要一个盛大的舞台，才能让每朵盛开的鲜花传播得更远！更美！「CREDAWARD 地产设计大奖·中国」就是地产设计界的一个盛大的舞台！

In an era of good time, a grand stage will make your talents shining and go publicity, and the CREDAWARD is the very stage to show your talents!

于 鹏 Steven Yu

立足公平、公正之本，挖掘中国当代优秀之作，「CREDAWARD 地产设计大奖·中国」将成为中国地产行业最值得尊重和期许的设计大奖！

Based on the foundation of fairness and justice, the CREDAWARD digs the outstanding contemporary works in China. The CREDAWARD will become the most respected and expected design award in real estate industry in China!

张春玲 Chunling Zhang

「CREDAWARD 地产设计大奖·中国」代表了中国地产行业最高水准的设计作品，作为国内地产首个国际性的以地产角度评选设计的大奖，在行业内具有很大的影响力和号召力，必将成为推动中国地产设计发展的重要力量，成为中国最具有公信力的地产行业设计大奖。在这里，不仅可以与同高水平的评委交流和探讨，了解和分享最新地产趋势、设计动向；同时，也可以了解更多优秀地产建筑师观点，在这里找到志同道合的伙伴、遇见优秀的设计公司以及优质的材料企业。

The CREDAWARD epresents the highest level of design works in real estate industry. As the first international design award in China, selected from the perspective of real estate, it has great influence and appeal in this industry, and will definitely become an important force to promote the development of real estate design in China, as well as, will become the most credible design award in real estate industry in China.Based on the platform set by the CREDAWARD, you will not only communicate and discuss with jurys who are all excellent but also understand and share the latest real estate and design trends; at the same time, you will also understand the various points from the excellent real estate architects, and find like-minded partners, as well as, meet excellent design companies and high-quality material enterprises.

张胜利 Shengli Zhang

祝「CREDAWARD 地产设计大奖·中国」越办越好，愿中国地产设计推陈出新，行业水准越来越高！

I hope the CREDAWARD will be better and better, and I also hope that the new works will emerge from the real estate design industry in China, and the industry standards will become higher and higher!

张兆强 Zhaoqiang Zhang

希望「CREDAWARD 地产设计大奖·中国」以高水平和公平、公正的态度，坚持成为助力中国地产设计发展的平台；同时，成为地产建筑师同行交流预判行业趋势，快速学习并了解各地产公司产品水平、设计机构设计能力的平台。担任评委期间，看到评委团队越来越壮大、参赛公司水平越来越高、奖项口碑在行业内的量级越来越高，影响力逐年扩大，我感到非常骄傲！

I hope the CREDAWARD, with a high-level, fair and just belief, sticks to become the platform to promote the development of real estate design in China as well as for the real estate architects to communicate and predict the industry trends, quickly learn and understand the product levels of the local companies and the design capabilities of the design institutions.During my time as a jury of CREDAWARD, I am very proud of seeing the team of jurys is growing, the level of participating companies is getting higher and higher, the magnitude of CREDAWARD is getting bigger and bigger in this industry, and its influence is getting expanded year by year.

朱晓涓 Xiaojuan Zhu

「CREDAWARD 地产设计大奖·中国」是中国独特、权威、公正的奖项，具备国际化的视野、专业水准及强大的社会影响力。每届参评作品的质和量是世界范围内罕见的，因为获奖难度极大，其已然成为优秀公司的角逐场，地产优秀作品的展示舞台。「CREDAWARD 地产设计大奖·中国」将竭力推动地产设计的交流和发展，成为建筑师、地产人专业的共享平台和窗口。

With an international perspective, professional standards, and strong social influence, the CREDAWARD is unique, authoritative, and fair. Thanks to the tremendous works involved in the annual event and the difficulty of winning the award, it has become the arena for excellent companies to compete and the stage for outstanding real estate works to be displayed. The CREDAWARD will strive to promote the communication and development of the real estate design and become the professional sharing platform and window for architects and real estate professionals.

祝峥 Zheng Zhu

今天的中国地产设计发展愈发成熟，消费也日益偏重品质，日趋理性，地产设计需要洞察需求，持续创新。愿越来越多的真正创造价值的设计作品通过「CREDAWARD 地产设计大奖·中国」这个广阔而具有权威性的发展平台，成为行业风向标，引领地产设计未来发展。

The current real estate design in China is becoming more and more mature, and the consumer is becoming more and more quality-oriented and rational, which require real estate design to look into insights and continuously innovate. I hope that more and more value-creating design works will become the industry benchmarks and lead the future development of the real estate design through the broad and authoritative platform of CREDAWARD.

嘉宾寄语
MESSAGE FROM GUESTS

·寄语排序方式：
按公司类别升序排列为地产、建筑、景观、室内、照明；同类别按嘉宾姓名首字母升序排列。

Yu Shi
时 宇

·融创中国控股有限公司
高级副总裁

祝愿「CREDAWARD 地产设计大奖·中国」成为中国地产领域的导航员，为行业发掘符合市场需求的产品，为客户创造有价值的好产品，好作品，导引行业不断向前。

Di Zhang
张 镝

·弘阳集团
副总裁

「CREDAWARD 地产设计大奖·中国」是业界标杆，它公平公正，立足于产品，关注客户感受，理解设计价值，被行业高度认可。希望大奖保持初心，促进行业健康有序发展！

Tong Zhao
赵 彤

·远洋集团控股有限公司
设计总监

远洋集团参加了历届「CREDAWARD 地产设计大奖·中国」，我们共同见证了大奖公正、公平、公开，深受业内认可及可持续的发展。大奖分类越来越精准，作品水平越来越高，涵盖面广，引领时代、促进行业发展，体现行业最高标准！通过大奖，我们也切实看到了地产行业设计水平的提升！愿「CREDAWARD 地产设计大奖·中国」持续引领行业前行！

Eric Phillips

·NBBJ 建筑设计事务所
合伙人 / 亚太区业务领导

通过参与「CREDAWARD 地产设计大奖·中国」，我们很荣幸能为我们的客户及其转型空间提供支持，以驱动创新、提升福祉并为健康社区提供支持。该奖项的认可证明了我们对建筑环境的影响是为了产生有意义的改变。

Xi Bo
薄 曦

·联创设计集团
董事长 / 总裁

这些年，中国各行业都产生了一些新品牌、新业务的领导者，它们随着中国经济的发展而涌现，随着各行业的发展和成熟而大放异彩。在中国设计行业，「CREDAWARD 地产设计大奖·中国」正是这样成长起来的，公平、公开、公正地打造成为中国地产设计中最具分量的奖项之一。在过去的几年里，「CREDAWARD 地产设计大奖·中国」和我们设计界同人相伴相生，更是成为我们行业作品被认可、被关注的最重要的展示平台。

Pu Chen
陈 璞

·筑土国际
总经理

我们从 2016 年开始每年参加「CREDAWARD 地产设计大奖·中国」，5 年来见证了奖项的成长，见证奖项越来越专业，在行业内的权威性不断提高，获得广泛认可。主办方不仅关注参赛作品，更是竭尽心力地对参赛公司及项目进行多元化宣传推广，为企业提升品牌价值提供了很好的平台。希望「CREDAWARD 地产设计大奖·中国」再接再厉，打造更具国际影响力的设计奖项，真正成为让世界瞩目的中国大奖。

程 蓉 *Rebecca Cheng*

- KPF 建筑设计事务所
  执行总监

KPF 自 20 世纪 90 年代进入中国以来完成了中国许多城市的地标项目，我们很高兴其中多个项目获得评委和大众的肯定并摘得了「CREDAWARD 地产设计大奖·中国」金银大奖。「CREDAWARD 地产设计大奖·中国」从建筑审美、环境影响、使用者体验和经济效益等多维度地评价地产项目给社会带来的积极影响，表彰关注建筑可持续性解决方案而做出努力的企业及机构，推动中国建筑行业的发展！希望大奖吸引更多优秀的海内外项目参与，不忘初心，为推动卓越建筑设计创建国际交流平台。

黄向明 *Xiangming Huang*

- TIANHUA 天华
  董事 / 总建筑师

「CREDAWARD 地产设计大奖·中国」是具有特殊意义的。在过去三十多年快速和剧烈的城市化浪潮中，地产对人民的生活和城市面貌变化起到了关键作用。曾几何时，我们看到人们住进了新房子而备感幸福，孩子们走进了新的学校而意气风发，我们也见证了一座购物中心改变了一座城市的生活品质。「CREDAWARD 地产设计大奖·中国」的评选把这些看似平凡但却深刻影响我们生活的巨量工作和我们对社会的责任、对设计的追求、对专业的责任连接了起来，让看似普通的工作有了更大的意义和更高的追求。建筑师同行们也有了更好的交流和沟通的平台。我衷心祝愿「CREDAWARD 地产设计大奖·中国」越办越好。

戴 烈 *Leon Dai*

- 建言建筑
  总裁

改革开放的中国正在进行史无前例的城市化进程，大规模的地产开发对建筑师来说既是机遇又是挑战。建筑设计要积累更要创新，「CREDAWARD 地产设计大奖·中国」为从业者提供了一个激励创新的平台。

江家旸 *Jiayang Jiang*

- 力夫设计
  合伙人 / 设计总监

不同于学术类设计奖项，「CREDAWARD 地产设计大奖·中国」聚焦于中国城市建设中的主体——地产项目，关注地产建筑师这个"普通"的设计师群体。它为行业甄选出无数经得住考验、引领设计潮流的优秀作品，为广大地产建筑师提供了一个思考设计、相互切磋的平台。毫无疑问，它是目前国内设计奖项中做得最好的：豪华的评审团、高手如云的参赛者、超高的含金量！未来「CREDAWARD 地产设计大奖·中国」定会在国际知名设计奖项中占有一席之地，引领国际设计潮流的发展！

胡劲松 *Jason Hu*

- 骏地设计
  总裁

"建筑，总是建筑师追求的学术理想和创作源泉。所有的建造，不应仅仅止步于为客户提供至高服务，而更应为了我们未来城市以更好、更可持续的发展提供服务。"
「CREDAWARD 地产设计大奖·中国」见证了城市与企业的成长。

李晓梅 *Eric Lee*

- Gensler
  大中华区执行总裁

有幸见证了「CREDAWARD 地产设计大奖·中国」从创立之初，到如今发展成为国际知名的地产行业大奖。大奖凭借其专业的评审角度、公正的评审流程，为建筑师 / 设计师们提供了一个展示作品和理念的绝佳平台，也为中国地产实践带来了具有国际视野的前沿观点。

Eric Lee
李颖悟

· 欧安地建筑设计公司
  创始人 / 总裁

祝贺「CREDAWARD 地产设计大奖·中国」推动地产创新的五年辉煌！祝愿在未来的新生活领域继续引领设计新趋势。

Jun Ren
任 军

· 天友设计集团
  首席建筑师

「CREDAWARD 地产设计大奖·中国」聚焦于中国城市化可持续发展，以创新和创意的价值观衡量地产行业，推动中国设计在技术和艺术层面的不断提升。

Zhongxiao Lu
陆钟骁

· 株式会社日建设计
  执行董事·全球设计总部中国区 总裁

日建设计作为参赛方有幸见证了「CREDAWARD 地产设计大奖·中国」这七年的发展历史。作为中国首个地产行业的国际性大奖，给投身设计行业的工作者提供了一个良好的平台，促进设计师之间、设计师与发展商之间增进合作和交流。期待「CREDAWARD 地产设计大奖·中国」在下一个5周年，能继续与我们设计方一同不辱使命，越来越多地亮相世界舞台，让设计真正成为艺术，使更多大众体会到设计的魅力。

Eric Shing
盛宇宏

· 汉森伯盛国际设计集团
  董事长 / 总建筑师

转眼间地建师评奖已举办了7年，幸喜于三年前获「CREDAWARD 地产设计大奖·中国」金奖，至今仍是勉励。以年鉴的方式将过往优秀项目做一次回顾和整理，对于行业沉淀极具历史意义。希望「CREDAWARD 地产设计大奖·中国」能始终保持这种高质量的甄选水平，敦促建筑设计行业的进一步发展与繁荣。

Qin Pang
庞 钦

· 贝诺 Benoy
  公司董事 / 上海公司负责人

作为「CREDAWARD 地产设计大奖·中国」"老"选手，我们逐年感受到了来自各个层级的设计力量的日渐蓬勃，打破了传统宏大和精英叙事的范围，这对整个中国的建筑设计生态的丰富和持续发展无疑促进巨大。祝越办越好！

Luoya Tu
涂洛雅

· 筑弧建筑设计（上海）事务所
  总裁

整理3年前的瀚海晴宇项目资料，至今仍然感慨万千。瀚海晴宇项目的入选是「CREDAWARD 地产设计大奖·中国」的评委们对团队工作的认可，更体现了市场各方对项目品质管理、成本管控上的高度肯定。书籍出版在即，衷心祝愿「CREDAWARD 地产设计大奖·中国」宏图大展，越办越红火。

Nan Wang
王 楠

- 德国施耐德 + 舒马赫建筑师事务所
  中国合伙人 / 设计总监

「CREDAWARD 地产设计大奖·中国」以其专业、专注的态度，为建筑师打造了一个全方位展示的平台，作为「CREDAWARD 地产设计大奖·中国」的老朋友，德国施耐德 + 舒马赫建筑师事务所将一如既往的支持、陪伴它的成长。

Wu Zhi Wei
吴志伟

- DP 建筑师事务所
  董事

「CREDAWARD 地产设计大奖·中国」创建了一个独特的平台，促进地产商、设计师及同业之间的对话和交流。它以地产的眼光审视设计的价值；以设计的眼光探讨地产的未来。其不断提升的专业性、开放性和国际性，将使其成为越来越有影响力的国际奖项。

May Wei
魏文梅

- CallisonRTKL
  副总裁 / 亚太区办公室负责人

亲眼看着地建师的「CREDAWARD 地产设计大奖·中国」一年年发展壮大起来，以严格的要求，公正的评选打造了中国地产界最具影响力的奖项之一，每年的奖项申请和评奖活动都是地产界和设计界的高光时刻，希望地产设计大奖继续传播表彰优秀设计理念，引领城市未来规划，提升国际国内设计标准！

Hua Zhang
张 华

- SUNLAY 三磊
  创始合伙人

希望通过「CREDAWARD 地产设计大奖·中国」的平台促进交流，深度思考中国建筑的未来走向，激发创造激情分享实践经验，创造更大价值。祝「CREDAWARD 地产设计大奖·中国」让建筑文化发扬光大，深入人心，越办越好！

Wei Wu
吴 蔚

- 德国 gmp 国际建筑设计有限公司
  合伙人兼执行总裁

"地产"不仅仅对中国经济，同时对于中国城市和建筑也是一个非常重要的词汇。它不仅改变了城市的面貌，也改变了人们的生活。不可否认，地产是趋利的，很多情况下会和公共利益相矛盾。但随着社会的发展，越来越多的地产公司、地产建筑师正在深度思考项目和社会的关系，如何在获取经济利益的同时，更好地完成社会责任。而这本书里呈现给读者的项目则是这些思考和实践的一部分，会给地产商启发，也给设计师信心和向往。希望「CREDAWARD 地产设计大奖·中国」坚持自己的理想和标准，为城市和人们的生活推荐更多的优秀范例。

Renzo Zhang
张润舟

- 汇张思建筑设计咨询（上海）有限公司
  董事长 / 首席执行官

很高兴我们曾获得第四届 2017-2018「CREDAWARD 地产设计大奖·中国」年度地产项目银奖，获奖项目万科南宁金域中央是基于人与自然和谐互动的理念，感谢评委团对未来趋势的引领，希望它也能成为未来生活的方式。

Dwight Law

· DLC 地茂景观
  创始人兼首席设计师

「CREDAWARD 地产设计大奖·中国」帮助公众了解何为优秀的城市设计。我们应该赞美城市及其复杂性，矛盾和局限性，并以一种永恒的设计语言促进城市的持续发展，基于当代文化行为模式，作为人类互动，娱乐和社区的催化剂。

刘志南 *Zhinan Liu*

· A&N 尚源景观
  首席执行官

「CREDAWARD 地产设计大奖·中国」为设计师、设计公司提供了一个获得认可、展示自我的平台。只有心中有梦，才能绽放更好的设计光辉，感谢「CREDAWARD 地产设计大奖·中国」。

李卉 *Hui Li*

· 纬图设计
  创始合伙人 / 设计总裁

非常荣幸能在「CREDAWARD 地产设计大奖·中国」2017、2018、2019 连续三年蝉联金奖，这是对纬图设计的肯定与鼓励。感谢组委会为行业设计师提供了一个促进交流、发展的平台，这将激励我们更加专注于设计，继续创造出更多优秀作品。

孙虎 *Eric Shing*

· 广州山水比德设计股份有限公司
  董事长 / 总经理 / 首席设计师

「CREDAWARD 地产设计大奖·中国」自创办以来，以敏锐的地产发展触觉和前瞻性的设计视野，不断发掘精品项目，引领设计趋势。期待通过平台的交流与影响力，持续推动地产设计的进步。

李方悦 *Clara Li*

· 奥雅设计
  奥雅设计联合创始人、董事、总裁 / 洛嘉儿童空间品牌创始人

洛嘉金山岭项目能够获得 2017—2018 年度的「CREDAWARD 地产设计大奖·中国」是对我们儿童友好型城市建设的重要肯定和支持！

谭伟业 *Philip Tam*

· 贝尔高林国际（香港）有限公司
  副董事总经理

「CREDAWARD 地产设计大奖·中国」为推动地产信息互融创造了优秀的交流平台，希望平台接下来能在分众化领域继续延伸，成为传播地产前端信息的支柱力量。

Chongling Ye
叶翀岭

- 朗道国际设计
  执行董事 / 设计总监

「CREDAWARD 地产设计大奖·中国」持续见证了地产设计行业的发展，提供了一个能够连接思想与智慧、思考与创造、理想与使命的平台，希望「CREDAWARD 地产设计大奖·中国」继续坚持精品与高质量发展，践行美好未来。

Kot Ge
葛亚曦

- LSD & 再造
  创始人

设计不是一种表达自我的演绎，而是一种克制自我、克制表达的尺度语言。正如「CREDAWARD 地产设计大奖·中国」坚持"设计拥有代表时代秩序的力量"。我们的设计，同样以肩负促进生活、城市发展乃至时代环境的美好使命为最高追求。基于这种共识，我们在参赛的过程中，感受到了组委会对设计的极大包容度与对整个市场和城市环境的品质追求，心怀感激。

Zhuo Zhang
张 灼

- 广亩景观
  执行董事 / 总经理

广亩景观运营十八年来一直对奖项申报不太热衷，直到我们持续观察了「CREDAWARD 地产设计大奖·中国」两三年，我们从尝试性参与到成为我们在国内参与项目最多、最重视的专业奖项，其原因很简单，即气场相合、三观一致，我们都是一群认真做事的人，我们都希望能够为行业带来正能量。希望「CREDAWARD 地产设计大奖·中国」继续为行业提供正向引导，推动行业健康发展！

Ricky Wong
黄志达

- RWD 黄志达设计
  创始人 / 董事长

「CREDAWARD 地产设计大奖·中国」作为国内极具影响力和号召力的奖项，为设计师们提供了一个非常好的展示平台。希望通过这个平台，激励业界设计精英们创造出更多优质项目及服务，促进行业共荣。

Jimmy Zhao
赵涤烽

- 安道设计
  执行总裁 / 董事 / 创始人

「CREDAWARD 地产设计大奖·中国」，作为地产行业中最具影响力和号召力的权威奖项，是引领和推动行业前行之驱动者。

Frank Jiang
姜 峰

- J&A 杰恩设计
  创始人 / 总设计师

中国城市的发展和生活方式的更迭，令未来的城市空间充满更多可能性，希望「CREDAWARD 地产设计大奖·中国」能更多关注并发掘充满可能性的空间，推动中国设计向更高水平发展。

Patrick Leung
梁景华

- PAL 设计事务所
  创办人 / 首席设计师

东方的设计真的越来越令人感动！在地产界的推动下，设计届上下齐心把水准愈推愈高，百花齐放，争艳斗丽，可喜可贺！可预计未来的发展必能影响环球！「CREDAWARD 地产设计大奖·中国」正是积极推动行业发展的重要平台之一。

Uno Lai
赖雨农

- 十聿照明设计有限公司
  董事长 / 设计总监

欣见「CREDAWARD 地产设计大奖·中国」这几年逐渐成为中国国内的指标性大奖，参赛作品愈来愈丰富且国际化。无论是参与或是得奖，对于参赛者来说都觉得荣幸，「CREDAWARD 地产设计大奖·中国」已成为各个优秀设计团队展现实力的平台。祝愿大奖每年都顺利举办，并且在未来愈办愈大，迈向国际！

Kris Lin
凌子达

- KLID 达观国际建筑设计事务所
  创始人 / 设计总监

「CREDAWARD 地产设计大奖·中国」为建筑师和设计师提供了一个专业的展示平台，并促进了地产行业高水准、创造力与创新的发展。

Jason Sun
孙嘉海

- 中奥光科（北京）国际照明设计有限公司
  总经理 / 设计总监

中国的就是世界的。「CREDAWARD 地产设计大奖·中国」对推动中国设计发展的意义深远。国际众多的设计大奖为推动全球的设计发展都起到了积极的作用，而在中国地产蓬勃发展并处于产业转型期的当下，立足于中国地产的设计大奖，将会激励所有的地产行业设计师更加深耕于本土项目，挖掘本土文化，为中国地产的发展贡献更多的智慧！

Haohan Chang
张颢翰

- 无界象国际设计
  创始人 / 执行总监

偶然机会下参加「CREDAWARD 地产设计大奖·中国」，应该说是我至今参加过最刺激的比赛，让设计者用 5 分钟阐述设计理念，而不是用图片评断作品好坏，设计师不就应该是会说会设计吗？很开心能获奖，这也是勉励我继续创作好作品的动力，希望能有更优秀的作品来参赛。

Xin Tian
田 鑫

- bpi
  执行董事兼中国区总经理

希望「CREDAWARD 地产设计大奖·中国」汇集中国地产设计之精萃，孕育中国地产设计之未来，引领世界地产设计之思潮。

| MESSAGE FROM GUESTS

· Sort by First Letter of Last Name

Chongling Ye

· LANDAU
  Executive Director / Design Director

The *CRED*AWARD has seen through the development of the real estate design industry, providing a platform to connect ideas and wisdom, thinking and creation, ideals and missions. We hope that the *CRED*AWARD will continue to adhere to high-quality development and achieve a bright future.

Dwight Law

· DLC
  Founder / Chief Designer

*CRED*AWARD helps educate to the public what good urban design should be. We should celebrate cities and their complexities, their contradictions and constraints and promote their continued evolution with a timeless design language based on contemporary patterns of cultural behavior as a catalyst for human interaction, recreation, and community.

Clara Li

· L&A Design
  L&A Chief Executive Officer / Co-founder
  Chief Inovation Officer / Principal

La V-onderland Chengde Jinshanling project won the prize of *CRED*AWARD (2017-2018), demonstrating the great recognition and support for our children-friendly city.

Eric Lee

· o.ffice for a.rchitecture + d.esign (OAD)
  Founder / President

Congratulations to *CRED*AWARD for the five-year brilliance in promoting real estate innovation! We wish the *CRED*AWARD will continue leading new design trends in future life field.

Di Zhang

· Redsun Properties Group Limited.
  Vice President

*CRED*AWARD enjoys the reputation of the industry benchmark thanks to its fairness and equity, caring works and clients, and understanding design value. I hope *CRED*AWARD to keep its original intention and make the industry develop healthily and orderly!

Eric Phillips

· NBBJ
  Partner / Asia Practice Leader

Through our participating in the *CRED*AWARD, we are honored to support our clients and their transformational spaces that drive innovation, improve wellness, and support healthy communities. The award recognition demonstrates the impact of our built environments to effect meaningful change.

Eric Shing

· Shing&Partners International Design Group
  Chairman / Chief Architect

The CREDAWARD has been held for 7 years. It is a great pleasure to win the Gold Award three years ago, which has been driving me all the time. It makes sense to review and sort out the previous excellent projects in the form of a book. We hope that CREDAWARD will continue to maintain this high-quality selection level and facilitate the architectural design industry to thrive.

Hu Sun

· Guangzhou S.P.I Design Co., Ltd.
  Board Chairman / Managing Director / Chief Designer

Since the establishment of CREDAWARD, with a keen sense of real estate development and a forward-looking design vision, it has been continuously exploring high-quality projects and leading the design trends. It is expected to take the influence of CREDAWARD to continuously promote the progress of real estate design.

Frank Jiang

· Jiang & associates creative design
  Founder / Chief Designer

The developing Chinese cities and changing lifestyles will provide many potentials for the future urban space. I hope that CREDAWARD will focus on and explore those potentials to lift Chinese design to a higher level.

Hua Zhang

· SUNLAY
  Founding Partner

I hope to take CREDAWARD platform to advance communication, think deeply about the trends of Chinese architecture, inspire creativity, share experience, and create greater value. I also hope CREDAWARD will let the architectural culture flourish and be deeply rooted in people's hearts as well as be better and better!

Haohan Chang

· MAXPRO Interior Design
  Founder / Executive Director

I happened to take part in CREDAWARD, which is the most exciting competition I have ever participated in till now. In terms of the rules of CREDAWARD, the designers take 5minutes to present their ideas instead of posting their drawings. It's great, isn't it? It's delighted to win, which also motivates me to create good works, bringing more competitive works to CREDAWARD.

Hui Li

· WTD
  ounding Partner / President

It is a great honor to win the Gold Award of CREDAWARD in the two consecutive years of 2017 and 2018, which encourages WTD, and makes us more confident. I would like to convey my thanks to the committee of CREDAWARD for providing a platform for the designers to exchange ideas and gain self-development. It is certain that we will focus on design and continue to create outstanding works.

Jason Hu

- JUND
  President

"Architecture is always the academic ideal and creative source that architects pursued. All construction should not only provide supreme services to the customers, but also for the better and more sustainable development of our future cities." The CREDAWARD has witnessed the growth of cities and companies.

Jimmy Zhao

- Antao
  Executive President / Director / Founder

The CREDAWARD, as the most influential and appealing authoritative award in the real estate industry, is the driver to lead and facilitate the industry.

Jason Sun

- IDDI Design Group, inc
  General Manager/ Design Director

What belongs to China belongs to the world. The CREDAWARD has a significant impact on the real estate design industry in China. Like many of the international design awards actively promote the global design industry, the CREDAWARD will do the same. At present when the real estate industry in China is booming and transforming, the CREDAWARD, based on China, will drive all the designers to be rooted in local projects, explore local culture, and contribute their talents to the development of the real estate industry in China!

Jun Ren

- Tenio Design
  Chief Architect

Focusing on the sustainable development of urbanization in China, the CREDAWARD measures the real estate industry with innovative and creative values. The goal of the award is to promote the continuous improvement of Chinese design in terms of technologies and arts.

Jiayang Jiang

- LIFE DESIGN
  Partner / Design Director

Unlike the academic design awards, the CREDAWARD focuses on the main body of China's urban constructions: the real estate architecture; focuses on the "ordinary" designer community of the real estate architects. The award has selected countless outstanding works that have stood the test and led the design trends for the industry; it has provided a platform for real estate architects to think about design and communicate. And there is no doubt that the CREDAWARD is currently the best in the domestic similar design awards due to its professional jury, top-notch participants, and high value. I believe that the CREDAWARD will be an internationally renowned design award and lead the development of international design trends!

Kot Ge

- LSD & Zaizao
  Founder

Design is not a deduction of self-expression, but a standard language that restrains oneself and expression. Just like CREDAWARD insists that "design has the power to represent the order of the times.", our designs also take the missions to advance life, to develop cities, to protect the environments as their highest pursuits. Based on that consensus, in the process of participating in the competition, we felt grateful for the organizing committee for their comprehension for the designs and the pursuit of the quality of the entire market and urban environments.

Kris Lin

· Kris Lin International Design
  Founder / Design Director

The CREDAWARD provides a professional exhibition platform for architects and designers. It also lifts the real estate industry to a higher standard with creativity and innovation.

May Wei

· CallisonRTKL
  Vice President / APAC Regional Office Leader

I see CREDAWARD growing year after year, which has become one of the most influential awards in the real estate industry in China due to its strict requirements and fair selection. Besides, it's a shining moment for the real estate and design industry to hold its annual award applications and activities. Hopefully, the CREDAWARD will continuously commend outstanding design concepts, lead the future planning of the city, and improve national and international design standards!

Leon Dai

· Verse Design
  President

The reform and opening-up China are undergoing an unprecedented urbanization process. The large-scale real estate development is both an opportunity and a challenge for architects. The architectural design needs to be accumulated and innovative. It is lucky that the CREDAWARD provides architects with an exciting and innovative platform.

Nan Wang

· schneider+schumacher International GmbH
  China Partner / Design Director

The CREDAWARD, with its professional and dedicated attitude, has created a comprehensive platform for architects to show their talents. As an old friend of CREDAWARD, Schneider+Schumacher will continue to support and accompany its growth.

Luoya Tu

· Archimorphic Inc
  President

I have compiled the data of HANHAI LUXURY CONDOMINIUMS, which was built three years ago, and I still have a lot of emotions. The selection of the project is the recognition of the teamwork by the Jury of CREDAWARD and it also reflects the high affirmation of all parties on project quality management and cost control. I sincerely hope that CREDAWARD will become more prosperous, and the book can come out soon.

Patrick Leung

· PAL DESIGN GROUP
  Founder / Principal Designer

It's amazing that the oriental design is more impressive! It is applauded that under the promotion of real estate industry, the designers and all the staff working for the design industry have made concerted efforts to advance the design industry to a higher level with inclusive and competitive works! It can be expected that the CREDAWARD will have a global influence in the future! It is CREDAWARD that is one of the important platforms to actively promote the development of the industry.

Philip Tam

- BELT COLLINS INTERNATIONAL (HK) LIMITED
  Deputy Managing Director

The CREDAWARD has created an excellent platform to promote the exchange of real estate information. I hope that the platform will continue to be extended in the field of segmentation in the future and become a pillar force for spreading the front-end real estate information.

Pu Chen

- Archiland
  CEO

We have participated in CREDAWARD every year since 2016. In the past 5 years, we have witnessed the growth of the award, witnessed that it has become more and more professional. The award's authority in the industry continues to increase and has been widely recognized. The organizing committee not only pays attention to the entries, but also makes effort to diversify the promotion of participating companies and projects, providing a good platform for companies to enhance their brand value. I hope that the CREDAWARD will continue to create a more internationally influential design award and become the Chinese award that attracts the world's attention.

Qin Pang

- Benoy
  Company Director/Head of Shanghai Studio

As Benoy has participated in past CREDAWARD, year by year, I feel the increasing strength of design from all levels, breaking the traditional grand and elite narrative, which undoubtedly contributes to the enrichment and sustainable development of the entire architectural design ecology in China. I wish CREDAWARD better and better!

Rebecca Cheng

- KPF
  Principal

KPF has built the landmarks in many cities in China since it entered China in the 1990s. We are very pleased to see that many of these milestone projects have been recognized by the jurys and the public and won the CREDAWARD gold and silver prizes. We understand that the CREDAWARD reviews the projects from multiple dimensions including architectural aesthetics, environmental impacts, user experience and economic benefits to exert a positive impact on society, commends the companies and institutions for their efforts to offer the solutions to building sustainability, and facilitates the development of China's construction industry! I hope the CREDAWARD will attract more outstanding domestic and overseas projects, stick to its original intention, and create an international platform to exchange ideas to promote the excellent architectural designs.

Renzo Zhang

- HZS DESIGN(SHANGHAI), LTD.
  Chairman / CEO

I am very glad that our project, Vanke Nanning Jinyu Central, has won the 4th CREDAWARD Silver Award for Real Estate Project. The project is based on the concept of harmonious interaction between human and nature. Thank the Jury for their guidance on future trends and hope the concept will become a future lifestyle.

Ricky Wong

- Ricky Wong Designers Ltd.
  Founder / Chairman

The CREDAWARD, as the most influential and prestigious award in the country, provides a very good platform for designers to display their talents. Through this platform, I hope to inspire the design elites to create more quality projects and services, promoting the prosperity of this industry.

Tong Zhao

· Sino-Ocean Group Holding Limited
  Design Director

Sino-Ocean Group has participated in the previous CREDAWARD. We have witnessed the fairness, fairness, openness and sustainable development of the award, which are well recognized by the industry. The award's categories are becoming more and more accurate, and the quality of works is getting higher and higher. CREDAWARD covers a wide range, leads the times, promotes the development and reflects the highest standards of the industry! Through the award, we have actually seen the improvement of the design level of the real estate industry! May CREDAWARD continue to lead the industry forward!

Xi Bo

· UDG
  Chairman / President

Like the leading new brands and new business emerged in various industries have been growing and getting stronger in China over the years, the CREDAWARD has been too. With fairness, openness and equity, the CREDAWARD has become one of the most important awards in real estate design in China. In the past few years, the CREDAWARD not only accompanied us but also became the most important platform to display the works recognized and highlighted by our industry.

Uno Lai

· Unolai Lighting Design& Associates
  Principal / CEO

It is delighted to see CREDAWARD has gradually become an indicative award in China in recent years, attracting the abundant entries with an international characteristic. No matter whether a participant or a winner, it is a great honor for them to be involved in CREDAWARD, which is the platform for excellent design teams to show their strength. I hope that CREDAWARD will be smoothly going well year after year, and be more influential, as well as, be global!

Xiangming Huang

· TIANHUA Group
  Board Member / Chief Architect

The CREDAWARD is featured with special meaning. In the rapid and huge wave of urbanization over the past three decades, real estate played a key role in the changes of people's lives and city's images. On the one hand, we see people happily moving in a new house, children merrily going to a new school, and city-dwellers pleasantly shopping at a mall. On the other hand, the CREDAWARD links those tremendous works, which are ordinary but influential in our lives, with our social responsibilities, our design dreams, and our professional responsibilities, making our ordinary work valuable, as well as, providing a better platform for architects to communicate. I sincerely hope the CREDAWARD will get better and better!

Wei Wu

· gmp International GmbH
  Partner / Managing Director

"Real estate" is not only an economically significant term but also a very important one when it comes to cities and buildings in China, because it changes the image of the city as well as people's lives. It is undeniable that real estate is profit-driven and contradicts public interests in many cases. However, with the development of society, more and more real estate companies and architects are thinking deeply about the relationship between projects and society, and how to better fulfill their social responsibilities while obtaining economic benefits. Thanks to this book, it provides readers with those thoughts and practices to inspire real estate developers and give designers confidence and hope. I hope that CREDAWARD will stick to its ideals and standards to recommend more excellent works for cities and people's lives.

Xiaomei Lee

· Gensler
  Managing Principal, Great China Region

It's my honor to witness the growth of CREDAWARD from the very beginning to the current internationally renowned award in the real estate industry. With its expertise and fairness, the CREDAWARD provides an excellent platform for architects/designers to display their works and ideas, and introduces globally edging views to the practice of China's real estate.

Xin Tian

- bpi
  Managing Director / General Manager of China District

I hope the *CRED*AWARD will pool the best designs of the real estate industry in China, nurturing its future, leading the trends of the global design in the real estate industry as well.

Wu Zhi Wei

- DP Architects Pte Ltd
  Director

he *CRED*AWARD has created a unique platform to promote dialogues and exchange of ideas among the real estate developers, designers and peers, letting them evaluate the value of designs from the perspective of real estate and explore the future of real estate from the perspective of design. The growing speed of its professionalism, openness and globalization will make it an influential international design award in the near future.

Yu Shi

- Sunac China Holdings Limited
  Senior Vice President

Hope *CRED*AWARD will be the navigator of the real estate industry in China, to discover good products and works that not only meet the requirements of markets but also create value for customers, to lead the industry as well.

Zhongxiao Lu

- NIKKEN SEKKEI LTD
  Executive Officer
  Principal, Architectural Design Group(China)

I am very glad that *CRED*AWARD is celebrating its 7th anniversary and NIKKEN SEKKEI has had the honor to witness the development history. As the first international design award in China, it provides a good platform for those who join in the design industry and promotes cooperation and exchange among designers and between designers and developers. At the same time, it is also expected that the *CRED*AWARD and our designers will continue to honor its mission, show up all over the world stage in the next 5 years. And I hope designing can truly become arts, and let more people experience the charm of designing.

Zhinan Liu

- A&N SHANGYUAN LANDSCAPE DESIGN
  CEO

The *CRED*AWARD provides designers and design companies with a platform for gaining recognition and self-expression. Only there is a dream in your heart can you shine better, please allow me to send my thanks to *CRED*AWARD.

Zhuo Zhang

- GM LANDSCAPE DESIGN
  Executive Director / General Manager

During the 18 years of GM Landscape, we have not been too keen on professional awards. One reason is that we feel our own accumulation not professional enough, and the other is that we feel difficult to meet a really suitable award to participate in. So, it wasn't until *CRED*AWARD running for several years that we really got involved. In the following years, our cooperation with *CRED*AWARD has become more and more intensive. And now it has become the most professional award in China. The reason is simple—we are a group of people who work hard and responsibly, we are all enterprises which bring positive energy with real hope to the industry. We hope that *CRED*AWARD can continue to provide positive guidance, and to promote the healthy development of the industry. We wish *CRED*AWARD better and better!

鸣 谢
THANKS

特别感谢各大地产公司、设计机构等伙伴多年来对 CREDAWARD 地产设计大奖·中国的参赛支持！
感谢全体评委多年来无薪酬公益参与奖项评审工作，致力于推动中国地产设计行业的创新发展！
感谢为本次图书出版共同参与审校工作的行业伙伴：（按姓名首字母排序）

边保阳　陈双霁　陈欣彤　程　悦　蔡越云　董　卉　邓　宁　高　毅　黄璟璐　黄丽玲　何梦婕　洪晓韵
黄一帆　金　春　贾　楠　贾娴静　蒋　毅　李华梅　林　静　李　明　李书娇　李暑霞　李　伟　李中伟
刘　和　林　欣　刘晓珊　凌云琪　凌　紫　马思媛　宁欣梅　阙加成　祁　希　申江海　孙巍巍　田　鑫
汤玲雁　唐聿炜　吴彩霞　吴　迪　王　婷　王蕴孜　吴志伟　谢白莎　许蒙娜　徐姗姗　杨存霞　叶蓁蓁
尹　荣　张　东　张恒豪　张凯宜　赵　爽　周　晟　郑珊珊　朱雯嫣　张　绛　左　秀　赵晓林　张晓庆
张秀媛　张　灼

Sincere thanks go to all the real estate companies, design agencies and our partners for their supporting CREDAWARD over the years!
Sincere thanks go to all the jurys, who reviewed the awards without pay through the years, for their devoting to facilitating the innovative development of the real estate design industry in China!
Sincere thanks go to the industry partners who participated in the review work for publishing this book:
(Sort by First Letter of Last Name)
Baoyang Bian · Shuangji Chen · Xintong Chen · Yue Cheng · Yueyun Cai · Hui Dong · Ning Deng · Yi Gao · Jinglu Huang · Liling Huang · Mengjie He · Xiaoyun Hong · Yifan Huang · Chun Jin · Nan Jia · Xianjing Jia · Yi Jiang · Huamei Li · Jing Lin · Ming Li · Shujiao Li · Shuxia Li · Wei Li · Zhongwei Li · He Liu · Xin Lin · Xiaoshan Liu · Yunqi Ling · Zi Ling · Siyuan Ma · Xinmei Ning · Jiacheng Que · Xi Qi · Jianghai Shen · Weiwei Sun · Xin Tian · Lingyan Tang · Yuwei Tang · Caixia Wu · Di Wu · Ting Wang · Yunzi Wang · Zhi Wei Wu · Baisha Xie · Mengna Xu · Shanshan Xu · Cunxia Yang · Zhenzhen Ye · Rong Yin · Dong Zhang · Henghao Zhang · Kaiyi Zhang · Shuang Zhao · Sheng Zhou · Shanshan Zheng · Wenyan Zhu · Jiang Zhang · Xiu Zuo · Xiaoling Zhao · Xiaoqing Zhang · Xiuyuan Zhang · Zhuo Zhang

图书在版编目（CIP）数据

中国地产设计集萃：CREDAWARD 地产设计大奖·中国 1-5 届金银奖项目．Ⅰ，2014—2019 / CREDAWARD 地产设计大奖·中国组委会编．-- 天津：天津大学出版社，2021.8
（中国地产设计发展史）
ISBN 978-7-5618-7011-2

Ⅰ.①中… Ⅱ.①C… Ⅲ.①建筑设计—作品集—中国—现代 Ⅳ.① TU206

中国版本图书馆 CIP 数据核字 (2021) 第 162335 号

ZHONGGUO DICHAN SHEJI JICUI Ⅰ (2014—2019)— CREDAWARD DICHAN SHEJI DAJIANG·ZHONGGUO 1-5 JIE JINYINJIANG XIANGMU

图书策划	地建师 | DJSER
责任编辑	姜　凯
组稿编辑	朱玉红
文字编辑	地建师 | DJSER
文字翻译	潘向翀
美术设计	高婧祎
图书制作	天津天大乙未文化传播有限公司

出版发行	天津大学出版社
地　　址	天津市卫津路 92 号天津大学内（邮编：300072）
电　　话	发行部 022-27403647
网　　址	www.tjupress.com.cn
印　　刷	廊坊市瑞德印刷有限公司
经　　销	全国各地新华书店
开　　本	260 mm×330 mm
印　　张	48.25
字　　数	400 千
版　　次	2021 年 8 月第 1 版
印　　次	2021 年 8 月第 1 次
定　　价	499.00 元

凡购本书，如有质量问题，请向我社发行部门联系调换
版权所有 侵权必究